New Dimensions in Science Fiction

Darwinian Feminism and Early Science Fiction

New Dimensions in Science Fiction

Series Editors
Professor Pawel Frelik
Maria Curie-Sklodowska University

Professor Patrick B. Sharp
California State University, Los Angeles

Editorial Board
Dr. Grace Dillon
Portland State University

Dr. Tanya Krzywinska
Falmouth University

Dr. Isiah Lavender III
Louisiana State University

Prof. Roger Luckhurst
Birkbeck University of London

Dr. John Rieder
University of Hawai'i

Darwinian Feminism and Early Science Fiction
Angels, Amazons and Women

Patrick B. Sharp

UNIVERSITY OF WALES PRESS
2018

© Patrick B. Sharp, 2018

All rights reserved. No part of this book may be reproduced in any material form (including photocopying or storing it in any medium by electronic means and whether or not transiently or incidentally to some other use of this publication) without the written permission of the copyright owner except in accordance with the provisions of the Copyright, Designs and Patents Act 1988. Applications for the copyright owner's written permission to reproduce any part of this publication should be addressed to the University of Wales Press, 10 Columbus Walk, Brigantine Place, Cardiff CF10 4UP.

www.uwp.co.uk

British Library CIP Data
A catalogue record for this book is available from the British Library.

ISBN   978-1-78683-229-0
eISBN   978-1-78683-230-6

The right of Patrick B. Sharp to be identified as author of this work has been asserted in accordance with sections 77 and 79 of the Copyright, Designs and Patents Act 1988.

Typeset by Marie Doherty
Printed by CPI Antony Rowe, Melksham

*For Sharon
and Tonks*

Series Editors' Preface

Science fiction (SF) is a global storytelling form of techno-scientific modernity which conveys distinct experiences with science, technology and society to a wide range of readers across centuries, continents and cultures. The New Dimensions in Science Fiction series aims to capture the dynamic, worldwide and media-spanning dimensions of SF storytelling and criticism by providing a venue for scholars from multiple disciplines to explore their ideas on the relations of science and society as expressed in SF.

Contents

Acknowledgements — xi

Introduction — 1

1 Scientific Masculinity and its Discontents — 13

2 Charles Darwin, Gender and the Colonial Imagination — 35

3 Evolution's Amazons: Colonialism, Captivity and Liberation in Feminist Science Fiction — 69

4 Women with Wings: Feminism, Evolution and the Rise of Magazine Science Fiction — 101

5 Darwinian Feminism and the Changing Field of Women's Science Fiction — 145

Works Cited — 175

Index — 189

Acknowledgements

This project began with a grant from the American Communities Program at Cal State LA back in 2009. Thanks to Maria Karafilis and my ACP colleagues for their support throughout the long road that led to the completion of this book. I am grateful to the Science Fiction Research Association for the research grant that helped fund a portion of my research on women science fiction authors of the 1920s and 1930s, and in particular I would like to thank Ritch Calvin, Susan George and Jason Ellis for their faith in this project. Thanks to the Office of the Provost and the Deans of Arts and Letters at Cal State LA for their support of the sabbatical that launched my work on the history of science and evolutionary discourse. Thanks to my colleagues with whom I collaborated at various stages of this project, especially Kimberly Hamlin, John Bruni, Sherryl Vint and Lisa Yaszek. I had the pleasure of working with some truly exceptional students during the process of researching this book, especially Joseph Aragon, Michelle Blackwell, Marissa Elliott-Baptiste and Mary Vasisth. I hope they learned as much from me as I did from them in the course of our independent studies together.

As with all recovery projects, I am deeply indebted to the amazing librarians – the true guardians of our past, present and future – with whom I worked as I tried to locate and examine the texts that are the focus of this book. Special thanks to Melissa Conway, Andy Sawyer and Daniel Lewis along with the staffs of the Eaton Collection at the University of California, Riverside, the Science Fiction Foundation Collection at the University of Liverpool Library, and the Dibner Collection at the Huntington Library.

Thanks to Sarah Lewis and the amazing people at the University of Wales Press for all of their efforts during the final stages of writing and revising this project, and to my co-editor Pawel Frelik for his hard work during the creation and development of this book series. Thanks also go to the peer reviewers for their thoughtful and precise feedback on the proposal and manuscript.

Thanks to Isiah Lavender III and the University Press of Mississippi for permission to reuse a few revised paragraphs from my essay 'Questing for an Indigenous Future' which first appeared in *Black and Brown Planets* (2014). Thanks also go to Rob Latham and Oxford University Press for permission to reuse several revised paragraphs from my essay 'Darwinism' which first appeared in *The Oxford Handbook of Science Fiction* (2014).

Special thanks to Sharon Sharp for all of her love and support during the process of researching and writing this book. And, as always, thanks to Savonia Sharp and Adelbert Sharp, to whom I owe everything.

Introduction

In Euro-American contexts, science fiction (SF) has long been a Darwinian genre. Of course, that is not *all* there is to SF – even SF in the Euro-American tradition – and much of this book argues that any such narrow understanding of the genre's essence is fatally flawed. Historically speaking, however, when the genre first began to be called 'scientifiction' by publisher Hugo Gernsback in his famous April 1926 editorial – a label that was later amended to 'science fiction' – the plots, characterisations and framing logics of the genre were based firmly within a Darwinian paradigm (Gunn 2005: ix; Sharp 2014: 475–8). This evolutionary paradigm has been contested, revised and reconfigured countless times in many disciplines since the nineteenth century, but it still provides the basic foundation for scientific understandings of all organic life. As such, it constitutes a major portion of 'the scientific megatext' and its attendant 'master narrative' that is understood, shared and sometimes contested by writers and readers of SF (Attebery 2002: 41). Scholars of SF have long recognised the importance of evolution for the scientific extrapolations of authors such as H. G. Wells, whose evolutionary parables about class and technological dependence provided a touchstone for twentieth-century science fictioneers. That said, Darwin's influence upon SF extended beyond such explicit scientific extrapolations: Darwin's work intervened in the pre-existing genre systems that shaped SF, bringing with it Victorian assumptions about colonisation, progress and human nature. Developing a critical understanding of Victorian evolutionary science is therefore necessary to appreciate the meanings and importance of twentieth-century SF.

SF is also a genre where women have made influential contributions throughout its history, however far back that history is extended. Building on the work of Carol A. Kolmerten, Jane Donawerth has traced 'a continuous literary tradition' from Margaret Cavendish's *The Blazing World* (1666) and Mary Shelley's *Frankenstein* (1818) to the explosion of women's SF in the 1960s and beyond (1997: xiv). Scholars such as Robin Roberts (1993), Justine Larbalestier (2002) and Lisa Yaszek (2008) have fleshed out major aspects of this tradition by respectively examining scientific discourse in SF by women, early SF magazine debates about gender, and the 'galactic suburbia' stories of the early Cold War. Anthologies such as Pamela Sargent's *Women of Wonder* (1975), Larbalestier's *Daughters of Earth* (2006) and Mike Ashley's *The Dreaming Sex* (2010) collected stories written by women and made them much more accessible to fans and scholars alike. *Sisters of Tomorrow* (2016), the anthology that I recently co-edited with Lisa Yaszek, shows the scope of women's contributions by collecting art, poetry, editorials, science writing and stories produced by women in magazine SF between 1926 and the mid-1940s. Together, such scholarship and anthologies have demonstrated beyond any reasonable doubt that women have always been important contributors and collaborators in the development of SF.

Like their male contemporaries, women science fictioneers of the early twentieth century wrestled with the meanings and possibilities provided by evolutionary science. However, many women SF writers adopted a more critical approach to evolution than their male counterparts, an approach that can be traced back to the work of feminists such as Mary E. Bradley Lane and Charlotte Perkins Gilman. The unique arguments and narratives produced by such Darwinian feminists provided an important foundation for SF written by women in the United States. Specifically, Darwin's account of sexual selection provided the template for diagnosing the violence of patriarchal institutions that enslaved women. It also provided an account of the possible mechanisms for their emancipation. Women such as Clare Winger Harris, Leslie F. Stone and C. L. Moore fashioned stories in the 1920s and 1930s that imagined how women could evolve and progress in ways that made them the equals – and at times, the superiors – of violent and egotistical men. As I show in the pages that follow, Darwinian feminism merged with the legacy of Margaret Cavendish and Mary Shelley in the writing of these women and provided scientifically enhanced futures that cast women as essential to progress.

Genre, taxonomy and the definitions of SF

Perhaps the most limiting aspect of SF scholarship over the past half century has been the contention that it is necessary to define SF – or the SF sub-genre being examined – in order for a scholarly project on the genre to have merit. The many attempts to define SF bring to mind the old analogy of educated men who are led into a dark room and asked to describe an elephant. One man describes the trunk, another the ear, a third the tusk, a fourth the tail, and so on. None of the men are wrong, but neither are they fully right. To more accurately describe the debates about SF, however, this analogy should be updated like this: several scholars are blindfolded and brought into a natural history museum's exhibit of the *proboscidea* (a taxonomic order that includes modern elephants) and asked to describe the animal they find there. One scholar describes the molar of a mastodon, another the ear of a woolly mammoth, a third the tusk of an Asian elephant and a fourth the tail of an African elephant. While the described body parts and animals are related to one another, they are characteristic of individuals and species that existed at different times and different places in the history and evolution of the *proboscidea*. Scholars discussing the history of SF have done something similar, assuming that all SF constitutes a discrete animal and attaching the genre to a fantastic tradition extending back to Homer's *Odyssey* (Gunn 1975), the Gothic legacy of Mary Shelley's *Frankenstein* (Aldiss 1988), the apocalyptic imagination of the Bible (Ketterer 1974) and the estranging narrative of Sir Thomas More's *Utopia* (Suvin 1979) among many, many other things. The attendant definitions of the genre have been equally diverse. These scholars were not wrong, but neither were they fully right.

So what, then, is SF? As several recent scholars have argued, this is probably the wrong question to ask (Kincaid 2003). That is because discussions of SF's characteristics and origins have been too wedded to models of genre that try to emulate scientific styles of taxonomy. Within SF studies, a paradigm shift over the past two decades has encouraged new ideas about genre that have changed our understanding of what SF is and how it emerged. In the 1960s and 1970s, many SF scholars produced definitions of SF that were based on formalist assumptions about genre. Their definitions provided the basis for measuring texts according to narrow, static criteria and excluding texts that did not measure up (Luckhurst 2005: 6–7; Rieder 2017: 7). Their goals were taxonomic: they were trying to come up with definitions of a literary genre that could be used to sort texts into categories, and in the process to identify the essence of a genre that was worthy of artistic praise and academic study. One major flaw with such formalist approaches to genre is that they try to pin down an essence that is constantly changing (and that does not exist at all). Genres are inherently dependent upon the recognition of patterns by authors, publishers, readers and everyone else involved in the production, circulation and reception of a text. Genres only exist across a vast number and range of subjectivities that do not always agree, which means that they are dispersed, contested and always subject to modification. Such a distributed and contested thing as a genre cannot be said to have any true essence.

So what can theories of genre tell us about culture, and what is a genre? Over the past few decades, scholars in the humanities and social sciences have developed accounts of genre that attempt to balance the analysis of formal textual properties with a more robust understanding of history and how genres evolve. *Genre* is the French word for 'kind' or 'type', and at a very general level a 'genre' is simply a kind or type of text. In a more specific sense, a genre is a repeated discursive form that is associated with (and that incorporates) particular social institutions, situations and ideologies. A genre is the overall form of what sociologists and anthropologists refer to as an *utterance*, which differs from the traditional unit of literary analysis called a *text*. Where *text* refers to a unit of discourse (e.g. a novel or a poem), *utterance* refers to an entire social situation that includes a text. In other words, when people analyse an utterance, they look at the social and material conditions under which a text is both produced and received. For sociologist Pierre Bourdieu, this means that a text is an occasion for a complex social negotiation between the producer(s) of the text, the receiver(s) of the text, and the various institutions and systems of power within which the producers and receivers are enmeshed. These negotiations and power structures are an integral part of the communicative act, and without them the text would be meaningless. This more complex and embodied social negotiation is the basic unit of analysis when trying to understand various forms of communication and representation. In this sense, a genre is not simply a set of formal textual properties; instead, it is a pattern of social negotiation that includes formal textual properties, but these properties are deployed and manipulated by producers for specific purposes, and they have definite material effects on receivers. This is the understanding of genre that literary critics such as Thomas O. Beebee have in mind when they say that 'a text's genre is its *use-value*' (1994: 14): genre analysis does not simply look for a formal pattern in a series of texts, but rather it tries to understand the meaning and purpose (or use value) of formal patterns in a particular historical moment. Following this approach, one way recent genre studies have avoided lapsing back into an unproductive taxonomic mode of analysis is to emphasise the 'social and historical aspects' of any given text (Miller 1994: 24). This approach takes up an understanding of genre as defined by Tzvetan Todorov, who asserted that a genre is 'the historically attested codification of discursive properties', and expands it to incorporate culture-wide understandings of narrative, discourse and form (1990: 19). Genres comprise a system that is constantly shaping, and that is shaped by, the ideology and institutions of a particular culture (18–19). Several recent theories of genre – in fields such as sociology, anthropology, linguistics, rhetoric, cultural studies, and film and television studies – have demonstrated that genres arise to solve recurring communicative or representational issues faced by members of a community (Bazerman 2004: 124–6; Bourdieu 1994: 129; Luckman 1992: 228). However, genres do not spring fully formed from the minds of authors. As Todorov describes, genres come 'from other genres. A new genre is always the transformation of an earlier one, or of several: by inversion, by displacement, by combination' (1990: 15). As a genre develops, certain formal elements are repeated and become codified because of their familiarity and success in handling communicative or representational issues.

Scholarship on the origins of SF has long emphasised this aspect of genre: SF was cobbled together from pre-existing genres such as the gothic romance, the extraordinary voyage, the tale of the future and the tale of science (otherwise known as the gadget story) (James 1994: 13–26; Kincaid 2003: 423). One limitation of late twentieth-century scholarship was the repeated attempt to locate SF within an 'academic-classical genre system' with an 'ancient lineage' that carried with it more 'cultural prestige' (Rieder 2017: 7). More recent scholarship, however, has emphasised why SF emerged when it did, the social needs and historical conditions the genre addressed and the entire system of genres that early SF authors drew from when writing their stories. In his influential cultural history of the genre, Roger Luckhurst cites such developments as mass education in science and literature, the impact of industrialism on everyday life and changes in the publishing industry as key 'conditions of emergence' for SF (2005: 16–29). John Rieder has argued that SF's 'precursor' genres – and SF itself – 'represent ideological ways of grasping the social consequences of colonialism' (2008: 20). As Rieder shows, and as I demonstrate in *Savage Perils* (2007), SF was heavily influenced by such closely related colonial genres as the scientific race treatise, the travelogue, the 'lost race' story and new narratives of human evolution.

In *Darwinian Feminism and Early Science Fiction*, I argue that the growing power of feminist movements and their embrace of evolutionary discourse in the late nineteenth century provided key conditions of emergence for women's SF. As historian Kimberly Hamlin demonstrates in *From Eve to Evolution* (2014), these developments fuelled feminist responses to Charles Darwin's account of human evolution in his *The Descent of Man, and Selection in Relation to Sex* (1871). Darwin's reframing of human sexuality proved particularly influential on American women involved in the struggles for women's suffrage and marriage reform. Darwin's account of sexual selection – and his essentialist account of the traits each human sex possesses – clearly were not feminist (Hamlin 2014: 1). However, as Hamlin notes, feminists 'were attracted to evolutionary science because, unlike biblical law, it was easily amendable and open to new ideas' (168). To counter the intractable story of Eve's curse, some feminists rooted their discussions in the supposed objectivity of scientific method, and appealed to Darwin's account of women's violent enslavement by men. Women such as Antoinette Brown Blackwell and Eliza Burt Gamble even put forward arguments for a scientific femininity that would complement the scientific masculinity entrenched within Euro-American scientific discourse and institutions. Darwinian feminists were ultimately marginalised from mainstream suffrage movements in the 1890s when the National American Women's Suffrage Association embraced a strange mixture of Christian conservatism and Spenserian progress (Hamlin 2014: 41–3). However, the writing of Darwinian feminists continued to find an audience, and novels such as Inez Haynes Gillmore's *Angel Island* (1914) and Charlotte Perkins Gilman's *Herland* (1915) captured the speculative dimension of Darwinian feminism that became common among the women publishing in SF magazines of the 1920s and 1930s.

One thorny issue when addressing the emergence of such traditions of writing is the degree to which the repeated structures of a genre become unconscious, and how specific ideologies become encoded and contested within these structures. In his essay 'The Whites of Their Eyes', Stuart Hall describes how a colonial 'grammar of race' became encoded in modern narratives in such a way that the structures of some texts 'have racist premises and propositions inscribed in them as a set of unquestioned assumptions' (1995: 20-1). As I argue in *Savage Perils*, SF authors of the early twentieth century such as Jack London and Edgar Rice Burroughs were quite conscious and explicit about their use of evolutionary discourse, racial 'science' and frontier adventure tropes in writing their stories. Over time, people working in many different media began to mimic and repeat the structures of London and Burroughs without being fully conscious of the ideological baggage these structures carried with them. For example, many overtly liberal, anti-nuclear, anti-racist stories written in the 1950s carried with them implicitly racist assumptions and plot structures about 'savagery' and 'civilisation' that the texts in other ways seem to be arguing against. This is an important issue when addressing the problem of SF's colonial origins and analysing how individual texts draw on and position themselves in relation to the genre's repeated formal elements.

Darwin's work contained not only a colonial grammar of race, but also a Victorian grammar of gender and sexuality. When writers such as London and Burroughs engaged in their evolutionary extrapolations, they included a series of assumptions about violent, intelligent, toolmaking men and nurturing, intuitive, fragile women as a part of their frontier adventures. Mark Feldman argues that the 'Darwinian body' in such narratives was 'also Gothic insofar as it was haunted by its past' wherein 'the subject replayed evolutionary history' (2010: 75). Feminist authors such as Gillmore and Gilman directly confronted this gothic Darwinian body, showing how horrible evolutionary history had been for women while imagining ways in which both women and men could evolve beyond the oppressive patterns of the past. At the same time, they embraced the formulation of women as cooperative and nurturing, asserting that these traits were the very ones that made civilisation possible in the face of violent Darwinian masculinity. It is difficult to know if women SF writers of the 1920s and 1930s had read Darwin's work, or even if they had read the work of Gillmore and Gilman. However, interviews and statements by women such as Harris, Stone and Moore make clear that they were strongly influenced by the writing of Burroughs and regularly read the work published in magazines such as *Amazing Stories* and *Wonder Stories*. In responding to the encoded Darwinian framework of the genre and the (masculinist) scientific megatext, these women – most of whom were highly educated – created characters and plots that closely resembled those of their Darwinian feminist predecessors. By creating such stories these women SF writers were creating 'heretical' texts, which is likely one reason their work has so often been excluded from the histories and anthologies of the genre produced after the Second World War. In his discussion of 'heretical discourse', Pierre Bourdieu points out the importance of representation in moments of perceived crisis:

> Heretical subversion exploits the possibility of changing the social world by changing the representation of this world which contributes to its reality or, more precisely, by counterposing a *paradoxical pre-vision*, a utopia, a project or programme, to the ordinary vision which apprehends the social world as a natural world. (1994: 128)

In doing so, a heretical text 'must not only help to sever the adherence to the world of common sense and ordinary order, it must also produce a new common sense and integrate within it the previously tacit or repressed practices and experiences of an entire group' (128–9). As Suvin famously argued, this kind of 'cognitive estrangement' is perhaps the most powerful characteristic of many SF texts (2005: 23–9). From Cavendish and Shelley to Harris, Lorraine and Stone, women authors used fantastic plots as they attempted to address the 'repressed practices' or needs of female readers in regard to the growing power of masculinist science and the scientific megatext. By altering their available genres – including the emerging genre of SF – in new ways, the 'heretical' texts by women SF writers can be understood as successful in addressing these needs because their plots, themes and character types became repeated and adapted in ways that constitute a unique tradition within SF and the larger body of women's writing in English that persists to the present day.

These women drew from familiar primary and secondary genres that were different from those of their male counterparts as they crafted their SF stories. In M. M. Bakhtin's formulation, 'primary' genres are 'simple' and 'unmediated', such as 'everyday dialogue' or letters (1986: 62). Secondary genres are 'complex' and generally constitute more 'highly developed and organised cultural communication' such as novels, scientific lab reports and political speeches (62). During their development, secondary genres generally 'absorb and digest' primary genres and give them a 'special character' (62). In this sense, it becomes necessary to talk about a text in terms of its *complex* of genres that are drawn from those primary and secondary genres available in a culture (Todorov 1990: 10). Within a given culture, the available genres constitute a system that serves as a resource for authors to tap into when creating a text, and that readers rely upon when reading a text. Women writing SF in the 1920s and 1930s were confronted with a scientific megatext and system of genres that routinely belittled or excluded them. As Jane Donawerth put it, 'from their patriarchal culture, later women writers of science fiction inherit[ed] science as a problem' (1997: xxi). However, like their Darwinian feminist predecessors, these women revised and reinterpreted the scientific megatext and the ideologies encoded within SF to create exciting adventures and progressive futures that accounted for their own experiences and desires. As Donawerth shows, their texts included developments in 'technologies of childbirth', 'scientific child rearing' and 'domestic duties' that freed women 'for further education and for public responsibilities' (14–15). In doing so, they often infused their work with evolutionary insights, elements of romance and domestic fiction, and everyday dialogue more familiar to women in their reimaginings of gender roles and sexual relationships. Drawing on such sources allowed them to place women in the centre of the action and provide pointed explorations of the benefits and dangers scientific masculinity might hold for women.

What was unique about much – but not all – of the SF writing by women in the late nineteenth century and early twentieth century was how they inserted these 'feminine' genre elements, female characters and feminist insights into plot structures and story types that were just as popular with male writers. Building on Donawerth's work, I have identified five frequent storytelling tactics that proved particularly popular among early women SF writers. The most central and important of these was the reorganisation of sexual selection, where women gain control of their bodies through biological, social and technological developments that are represented as positive progress. This includes control over their romantic destinies and the freedom to choose a partner who best suits their temperaments and respects their rights and abilities, or indeed the right to choose no partner at all. Implicitly or explicitly, these stories champion women's control of reproduction for the betterment of the species and eliminate the threat of rape, dangers of childbirth and burdens of child rearing. The other four narrative tactics were largely put into the service of this first one.

The second tactic employed by women SF writers was the fantastic expansion of the domestic sphere. As Lilith Lorraine stated in 'Into the 28th Century' (1930), this utopian tactic imagined what would happen if women 'enlarged the scope of maternity and the four walls of her home to include the spiritual and intellectual guidance of our planet, the home of the human race' (122). The most well-known predecessor for this was Gilman's *Herland* (1915), where a nation run by women domesticated an isolated landscape and built a perfect society free from masculinist oppression. Women writing SF in the late 1920s and 1930s imagined domesticating nations and planets, and spun intergalactic and interdimensional adventures where women expand their civilised vision of home through conquering the brutal savagery of men. This tactic often incorporated elements from 'feminine' genres such as domestic fiction in order to create futures that were more appealing to women and that more fully addressed their needs.

This expansion of the domestic sphere often went hand in hand with a third storytelling tactic, one where women SF writers explicitly gendered and reconfigured the colonial gaze so that women are seen as the carriers of civilisation who must overcome the savagery, arrogance and narrowmindedness of men. As John Rieder explains, 'the colonial gaze distributes knowledge and power to the subject who looks, while denying or minimizing access to power for its object, the one looked at' (2008: 7). One of the most persistent powers of SF as a genre is its propensity for inverting or reconfiguring the colonial gaze in a way that denaturalises dominant power structures. In the hands of Darwinian feminists, inverting or reconfiguring the colonial gaze became a clear way to highlight the oppressed status of women in American society, as well as a playful way to imagine what women might do if they gained hegemony over men. Such stories took Darwin at his word when, in *The Descent of Man*, he associated sympathy with women and held it up as the highest civilised virtue. Civilised women thus became the heroes of these technofutures as they expanded feminine forms of progress across globes, galaxies and dimensions. In this way, these women developed a feminine form of colonisation that sometimes contested and sometimes reinforced the hierarchies of masculinist Euro-American science and Euro-American colonial rhetoric.

A fourth storytelling tactic commonly employed by women SF authors was to warn of the dangers of masculinist science. The immense influence of Mary Shelley's *Frankenstein* affected the writing of both women and men during this period, and tales of mad scientists were ubiquitous in all of the magazines that published SF. However, the women SF writers I refer to as Darwinian feminists distinguished themselves from their male counterparts by dramatising the unique dangers science posed to women, and by developing female characters who served as the primary voices warning about a masculinist science that attempts to dominate or warp a nature that is coded as female. More specifically, Darwinian feminists created stories where mad men of science tamper with the mechanisms of evolution, and in the process attempt to force women into sexual relationships or unnatural modes of reproduction against their will. Women SF authors often created an alternative scientific femininity that was more nurturing and ethical than traditional scientific masculinity, with women super scientists bringing new kinds of progress to the universe. They also regularly called into question the colonial enterprises enabled by masculinist science, representing conquest and plunder as vile and savage products of Darwinian masculinity.

The fifth common storytelling tactic embraced by women SF writers of the period was to represent Amazons and angels as the apex of Darwinian feminist evolution. The image of the Amazon provided an obvious role model from classical culture for feminists, but the colonial context of evolutionary discourse allowed for Amazons to serve as alternative models of evolutionary development freed from the restraints of masculinist Euro-American society. In a cultural moment where Amelia Earhart was the most salient feminist icon in relation to scientific and technological progress, women SF writers also imagined a host of flying women as exemplars of liberated Amazonian femininity. Whether the wings were represented as biological products of eugenics or as technological extensions of the organic body, wings allowed Darwinian feminist heroines to fly through their worlds in ways that took them far from the domestic sphere. Such stories generally provided a stark contrast to those that fantastically expanded domesticity: for Amazons and angels, the domestic sphere was a prison from which they could finally escape in pursuit of the full expression of their potential.

Chapters one and two of this book briefly trace the origins and development of Euro-American scientific masculinities, especially as they took shape in the colonialist narratives of men such as Francis Bacon and Charles Darwin. In the process of surveying these narratives, chapters one and two also delve into the responses of women – especially feminists – who pointed out the limitations and possibilities of scientific discourse and began to employ the five storytelling tactics discussed above. Chapter three analyses the impact of Darwinian feminism on the utopian tradition of women's writing in the late nineteenth century through the 1910s. Chapter four is the longest, as it seeks to recover the significant contributions women writers made to early magazine SF of the late 1920s and early 1930s. Chapter four places a particular emphasis on Darwinian feminism's influence in the storytelling of the women authors from this period, and shows how some women began to reconfigure or critique Darwinian feminism's tenets as they fused it with the traditions of Shelley and the new subgenres of Gernsback-era SF. Chapter five concludes the book, and shows how changes in SF magazine editing in the mid-1930s contributed to the eclipse of Darwinian feminism and the rise of women who published SF that was often indistinguishable from their male colleagues. This chapter also briefly surveys the writing of women in the 1920s and 1930s whose work does not fit within a Darwinian feminist rubric, and ends with a discussion of where Darwinian feminist SF fits within the long history of women's SF.

1 Scientific Masculinity and its Discontents

There is a great deal SF scholars can learn from the critical study of science by scholars in the humanities and social sciences. Unfortunately, as Roger Luckhurst put it over a decade ago, 'the strangest silence in SF scholarship has surely been the marginal interface between SF critics and those in Science and Technology Studies and History of Science programs' (2006: 2). Recent studies in the history of science have shown how early modern science – and what is known as the Scientific Revolution – was intertwined with European colonial expansion. In the introduction to the 2000 issue of *Osiris* entitled *Nature and Empire: Science and the Colonial Enterprise*, editor Roy MacLeod notes that for historians of science, the 'scene has shifted considerably from asking whether science was a feature of imperialism (it was, and is), and whether imperialism advanced science (likely, if not always possible to prove), to a broader range of questions' (2000: 11). Some recent scholarship on SF has moved in a similar direction. SF's obsession with the frontier and meditations upon racial difference are now understood to be two key markers of the genre's colonial origins. However, little work has been done to chart how both scientific narratives and SF drew from – and fed back into – larger colonial systems of genres as they were taking shape. Indeed, the failure of SF scholars to engage with critical scholarship on the sciences continues to be a major limitation of the field. Despite the recent work of scholars such as Luckhurst, Sherryl Vint and Colin Milburn – and the important foundational work of feminist scholars such as Donna Haraway, Constance Penley and Robin Roberts – many scholars remain uncritical of the 'science' part of SF studies.

What follows for the next two chapters is a brief history of Euro-American scientific narrative that pays particular attention to narratives of colonisation and the ways in which scientific masculinity and gendered formulations of nature became woven into the 'scientific megatext' (Attebery 2002: 41). Erika Lorraine Milam and Robert A. Nye note that it was 'predominantly male groups that shaped the work of science, technology, and medicine' and 'authorized the construction of gendered and sexed bodies' (2015: 2). In the process of defining gender and sex, these male groups drew on mutually constituting 'masculine/feminine binaries' that excluded women from 'scientific cultures' along with 'qualified men ... who did not seem to be the "right" kind of man' (3). In this sense, perceived differences in class, religion, sexuality and eventually race became involved in 'methods of differentiating between kinds of men' (3). Masculinity was never monolithic, and at different times and places scientists were able to 'choose from among a variety of masculine roles, including laboratory-based scientist-heroes, outdoor, self-reliant men, sensitive and sympathetic readers of nature, and family men' (5). These masculine roles in scientific cultures became a basis for fictional and popular understandings and critiques of scientists and their various projects. Gendered formulations of scientific identity went hand in hand with narratives of colonial domination, and became central to the nascent speculations we now call SF. They also provided the hegemonic visions against which the first women writers of SF struggled as they spun heretical tales of heroic female scientists and unscrupulous men alienated from feminine morality.

Narratives of discovery

In the late 1400s and early 1500s, the printing press allowed accounts of the voyages of Christopher Columbus and Amerigo Vespucci to receive a wide distribution in Europe (Loewen 2007: 38). A key feature of this new genre of discovery narrative was the detailed description of the 'discovered' lands and natives from the perspective of the male explorer. These early discovery narratives included detailed descriptions because of their connection to the colonial and mercantile purposes of such voyages: the authors were cataloguing the various resources of the lands they came across to make clear (and at times exaggerate) the profits and power that could be gained by future expeditions. If the folks back home who were footing the bill liked what they read, then it would be easier to get more funding down the road. These narratives often appealed to the Christianising mission of educating the natives and saving them from a chaotic existence that included cannibalism. In essence, some of these narratives suggested, Europeans were doing the natives a favour by bringing them under the umbrella of civilised European order. Not all colonial discovery narratives were flattering. Critics such as Bartolomé de Las Casas, a Dominican friar, emphasised in their accounts of colonisation the extreme cruelty and violence of the Europeans. However, a large number of people came to hail Columbus, Vespucci and scores of other explorers as heroes of Christianity, civilisation and progress (Berkhofer 1979: 5–12; Thomas 2000: 3–10; Zinn 2005: 1–12).

Whether condemning European depravity or providing glowing accounts of their journeys, these early gendered narratives of discovery became prominent within the system of genres circulating throughout Europe by the seventeenth century. An early fictional voyage that drew on this emerging genre was Francis Godwin's *The Man in the Moone* (1638). In the preface to a 2009 scholarly edition, William Poole describes Godwin's text as the 'first work of English science fiction that can claim some title to that status' (Godwin 2009: 7). Though some might consider Francis Bacon's 1627 utopia *The New Atlantis* (discussed in the next section) as an earlier English-language SF narrative, Godwin's text shows explicitly the connection between gender, colonial narratives and techno-scientific progress in the anglophone imagination of the seventeenth century. The story is narrated by Godwin's diminutive Spanish protagonist Domingo Gonsales. After leaving school to pursue some adventures, the Spanish nobleman finds passage to the East Indies with '2000 Ducats' where he is successful in his trading, earning a 'yeeld ten for one' on his investment (2009: 74). Falling sick on his return voyage, however, he is left on an idyllic island named St Helena to recover with only 'a *Negro*' named Diego to attend to him (76). While recovering on the island, Domingo muses, 'I cannot but wonder, that our King in his wisdome hath not thought fit to plant a Colony' on St Helena (74). Godwin's story rationalises Domingo's adventure in exploitive colonial terms, emphasising the financial value of such voyages and reinforcing emerging European colonial hierarchies by giving him a black servant. Godwin makes clear the financial incentive to plant colonies, as Domingo's desire to put a colony on St Helena is justified by its ideal location for servicing ships travelling from the East Indies to Europe.

Godwin's work also reveals the connection between colonial exploitation and the emerging techno-scientific desire to control nature. Using his year on the island to master the local animals, Domingo trains some 'wilde *Swans*' he calls '*Gansas*' and creates a rig that allows the birds to carry his small frame through the air (76). When he finally takes his voyage home on a Spanish ship, superior British ships attack and Domingo is forced to save himself: he takes out his '*Gansas*' and attaches them to his 'Engine', trusting that they 'for safeguard of their own lives (which nature hath taught every living creature to preserve to their power) would make towards the Land' (83–4). As a master of nature, Domingo is able to harness the very survival instincts of animals for his own purposes. However, the island his birds land upon is the site of 'continuall warre' between Spanish colonisers and 'a Savage kinde of people' (85). When the 'Savages' see Domingo land, they come down from the hills to attack. This dramatic aspect of colonial adventure – where inferior and bloodthirsty indigenous people threaten the techno-scientific hero of civilisation – would become increasingly important for both non-fiction and fictional narratives of male scientific prowess. Through his use of superior technology and his mastery of nature, Godwin's lone inventor genius outwits a horde of natives, thus proving his own superiority and the superiority of his civilisation.

This is where the narrative turns to the fantastic, and where Godwin builds a direct link between colonial adventure on earth and a voyage into outer space that would become repeated in the later works of Jules Verne, Edgar Rice Burroughs and countless others. When Domingo's birds flee the natives, they take him on their migratory route straight up, eventually landing on 'the New World of the Moone' (97). There he meets a society of giants superior in every way to the people on earth. The description of the 'Lunars' and their society is consistent with travelogues and anthropological accounts of the period (99, n. 4). Where travelogues and fantastic voyages of the eighteenth and nineteenth centuries increasingly emphasised physical (and especially racial) differences, Godwin's text focuses more on clothing and religion. The superiority of the Lunars is epitomised by their hatred of vice and their embrace of a fervent Christianity. The landscape of the moon is depicted as Edenic, and Lunar culture is described in utopian terms. However, Godwin also makes clear that the natural colonial order of earth extends to the moon: the Lunar people have developed a science whereby they can tell 'by the stature, and some other notes they have, who are likely to bee of a wicked or imperfect disposition' (113). In essence, Godwin promotes here an early version of eugenics, and those 'wicked or imperfect' people are banished to the earth. In another colonial twist, the people they banish are identified as the people of the Americas. In this way, Godwin's narrative shows early traces of the colonial scientific racism that would become central to discovery narratives and travelogues over the next two centuries. Though he uses the Lunars to criticise the corruptions of European culture, he also strongly reinforces the belief that Europeans are the superior people of earth, and that European techno-scientific prowess is the key to expanding their dominion and horizons across the globe and into space.

Colonialism, gender and scientific masculinity

Godwin was not alone in linking colonisation, science and masculinity as central to progress. Narratives created about the 'new science' by writers such as Francis Bacon included accounts of male scientists subduing a female nature. As Mary Terrall shows, 'The "new science" of the seventeenth century has long been linked to the voyages of discovery that expanded the conceptual and physical horizons of the European world' (1998: 226). Bacon explicitly used the voyages of discovery as an inspiration for discarding older ways of thinking about nature and pursuing new paths to knowledge. A recurring image of discovery in the sixteenth century represented colonised lands such as the Americas as virginal females tempting or yielding to the male explorer (Brookes 2006). Explorers such as Sir Walter Raleigh represented 'territorial conquest as the enforced defloration and possession of a female body' (Montrose 1991: 30). Bacon kept the perspective of the male explorer, and altered the object of his exploration from a virginal female to a deceitful witch who reveals the truth under forceful interrogation (Merchant 1990: 168-9). Like Raleigh, Bacon's imagery highlighted the domination of males over females.

As with the colonial voyages of discovery, the benefits the new Baconian science offered included wealth, power and prestige. However, as Katherine Park argues, the 'Scientific Revolution' was not the gloriously positive development traditional histories of science made it out to be: 'Instead of liberating the human mind and laying the foundations for general human happiness, it both reflected and encouraged the continued and increasing subjection of women and the exploitation of the natural world' (Park 2006: 490). A common feature of the colonial scientific gaze as it developed in the seventeenth century was the movement away from the image of nature as female toward the mechanical conception of nature as a machine. The mechanical view of nature (or Mechanism) was elaborated by French thinkers such as René Descartes and Pierre Gassendi, as well as by Englishmen such as Thomas Hobbes. As Merchant shows in her landmark work *The Death of Nature*, Mechanism 'laid the foundation for a new synthesis of the cosmos, society, and the human being, construed as ordered systems of mechanical parts subject to governance by law and to predictability through deductive reasoning' (1990: 214). In response to the religious, political and social upheavals of the seventeenth century, Mechanism provided an image of nature that was not unruly; rather, nature was inert and subject to control by those who carefully studied its laws. Likewise, social problems could be solved by the exercise of reason and authority. Mechanism promised to bring order to the chaos of European societies, and – through the ongoing enterprises of colonisation – to bring order to the rest of the world. The machine became 'an image of the power of technology to order human life' (Merchant 1990: 220).

Since this 'Scientific Revolution', modern scientific activity has been represented as a masculine pursuit in Euro-American culture. However, before the modern scientific academies were established in the late seventeenth century, women with social access participated in the scientific activities of artisan workshops, salons and royal courts (Noble 1992: 197–204; Schiebinger 1989: 17–19). These social and physical spaces were important in the development of the new empirical approaches of natural philosophy, approaches that rejected the teachings of the ancients dominating university curricula at the time. With few exceptions, the universities were closed to women; as such, the development of this new science outside universities provided women with important opportunities to contribute to knowledge about the natural world. This was particularly important in England, where Henry VIII's closure of convents during the 1530s and 1540s eliminated the major centre of 'spiritual and intellectual life' for women (Schiebinger 1989: 13). In villages and among the poor, medical treatment usually came from 'wise women' such as midwives who practised a 'popular magic' learned from oral traditions. Women even played a major role in the development of the hermetic tradition of science in the seventeenth century (Noble 1992: 187–8).

The establishment of institutions such as the Royal Society of London marked both a legitimisation of the 'new science' as well as 'the formal exclusion of women from science' (Schiebinger 1989: 20). There were a number of factors that contributed to this formal exclusion. In medieval Europe, the pursuit of knowledge had become associated with a denial of the pleasures of the flesh. In particular, celibacy became a defining feature of the Christian 'clerical asceticism' that dominated academic life in monasteries, convents and universities. When the Royal Society was founded in 1660 (it was given a royal charter in 1662), members such as Robert Boyle championed 'the celibate ideal' in spite of the fact that they lived in 'Protestant, anti-monastic England' (Noble 1992: 226). For many of its male members, excluding women was an obvious step to take to ensure the respectability and purity of the institution (Noble 1992: 225-9; Schiebinger 1989: 12, 151-2).

The English Civil Wars and the Interregnum of the 1640s and 1650s also had a major effect on the exclusion of women from the Royal Society. The 1640s and 1650s saw the growth of various forms of religious dissent, political radicalism and social upheaval. With the Restoration of Charles II to the throne in 1660, there was a powerful backlash against the unrest and radicalism of the previous two decades. This conservative backlash took many forms, but one focus of conservative energy was to attack the freedoms and positions of power women had gained over the past two decades. In the late 1650s, future members of the Royal Society began to distance themselves from hermetic beliefs they considered to be 'feminine' and aligned themselves with a politically neutral vision of science. The Royal Society was thus instituted as a home for a science that was 'masculine' and in opposition to radical and 'feminine' forms of pursuing knowledge (Keller 1985: 45-7, 51-4, 62-3; Noble 1992: 185-9, 228-9).

The writing of Francis Bacon from the early 1600s played an influential role in the Royal Society's move toward a 'masculine' science. In his writings such as 'The Masculine Birth of Time' (1603) and *The New Atlantis* (1624), Bacon put forward a vision of a 'science and technology ... with the power to transform man's relation to nature' (Keller 1985: 48). The narrative of scientific activity that emerged from Bacon's writing emphasised a male scientist pursuing a female nature to extract her secrets. This is not surprising: as Londa Schiebinger observes, 'From ancient times to modern times, nature – the object of scientific study – has been conceived as unquestionably female. At the same time, it is abundantly clear that practitioners of science – scientists themselves – have overwhelmingly been men' (1989: 122). However, the language that Bacon used seemed to call for a new relationship between the male scientific investigator and nature. As Carolyn Merchant has repeatedly shown, Bacon used metaphors of torture to characterise the new scientific approach for which he advocated (2006: 518–29; 2008: 733–5). In Bacon's writing, the scientist was in the position of the inquisitor trying to wrest secrets from a witch who was associated with nature.

In the utopian vision of *The New Atlantis*, Bacon provided a model institution that was one inspiration for the Royal Society decades later. *The New Atlantis* imagined a rigidly patriarchal society epitomised by the (male) scientists working in a research centre known as Salomon's House. Bacon's account of scientific progress entailed extending the control and empire of scientific men over the entire natural world (Merchant 1990: 172–6; Noble 1992: 223–4). Baconian imagery and language played a role in the ongoing struggle against hermetic visions of the new science in the 1650s and 1660s. The hermetic tradition of science saw nature in hermaphroditic terms, with male and female principles operating throughout nature. Some future members of the Royal Society attacked hermetic thought as 'feminine' and sensual. After the founding of the Royal Society, some members continued to attack proponents of hermetic philosophy and associated them with witches (Keller 1985: 50–61; Noble 1992: 187–8). The Royal Society embraced a conservative masculine ideal that saw nature as an unruly female that was to be controlled by men, just as Bacon had envisioned. Though rejecting the political power for scientists that Bacon envisioned in *The New Atlantis*, the Royal Society moved forward with a clear ideal of controlled and controlling scientific masculinity. Feminists of the early modern period put forward a number of arguments for the equality or superiority of women, particularly in regard to their fitness to participate in scholarly pursuits. However, women became increasingly marginalised from the centres of the new science (Schiebinger 1989: 165–70).

Margaret Cavendish and her *Blazing World*

The work of Margaret Lucas Cavendish, duchess of Newcastle-upon-Tyne, provided perhaps the earliest example in English of a woman author who critically examined the connections between gender, European colonial expansion and the imaginary framework of modern science. The upheavals of the British Civil Wars that shaped the founding of the Royal Society also shaped the life of Cavendish, a royalist whose husband lost a great deal of property and wealth due to his allegiance to the Crown. However, Cavendish 'criticized mechanical and experimental philosophy' and provided a singularly important voice that was contrary to the Royal Society (Sarasohn 2010: 2). In the consolidation of the Royal Society as an exclusively masculine space, Cavendish was the woman who was most conspicuously excluded: she was only allowed to visit a meeting of the society once on 30 May 1667, and was never allowed to join despite her well-known scientific writings (Hutton 2003: 161–2; Sarasohn 2010: 29–33). This exclusion helped fuel Cavendish's playful use of genres in the 1666 publication of a scientific text, *Observations Upon Experimental Philosophy*, which was coupled with the fictional adventure *The Description of a New World, Called The Blazing World* in the same volume. It also fuelled her satirical description of scientific institutions in *Blazing World*, a text that was published the year before her visit to the Royal Society. In her publications – and during her visit to the Royal Society – Cavendish displayed a keen awareness of gendered performance. Cavendish 'incorporated elements of male clothing' into her dress for her visit to the Royal Society, thereby 'problematizing her gender' and tweaking the conservative ethos of many of the society's members (Sarasohn 2010: 27–8). This 'hermaphroditic' performance of gender was 'a reenactment of the spectacle her other self in the Blazing World had already performed' (27, 33). In her fiction and her lived performances Cavendish engaged in a celebration of science and a critique of scientific masculinity that would become commonplace in feminist SF of the twentieth century.

Robert Boyle was an influential proponent of the new kind of scientific masculinity that became an object of Cavendish's scorn. In Boyle's vision of this masculine ideal, a good Christian scientific gentleman demonstrates that he is above the temptations of his social position and adopts the 'posture of the disengaged searcher after truth' (Shapin 1994: 150). Including elements such as 'physical frailty as a badge of spirituality', Boyle's ideas and lived example formed the framework for a scientific masculine ideal that still circulates in the genre system of American culture (Shapin 1994: 152). The pomp and circumstance of Cavendish's visit to the Royal Society upstaged the staid scientific performance of Boyle, who presented a number of experiments for the occasion. Cavendish's statements of admiration for the experiments were likely another caustic part of her performance: she had already made clear in her writings that the Baconian experimental tradition embraced by the Royal Society was 'deluding' and without benefit (Hutton 2003: 165–6; Sarasohn 2010: 30–3). Through her written and embodied performances, Cavendish used satirical humour to undercut the pretensions of scientific masculinity and draw attention to its limitations.

The content and structure of *Blazing World* also drew critical attention to the colonial contexts and rhetorics that shaped the new science embraced by the Royal Society. In an introductory poem to *The Blazing World*, Margaret Cavendish's husband William, the first duke of Newcastle, puts her imaginary voyage in a familiar colonial context. He notes that,

> Columbus then for Navigation fam'd,
> Found a new World, America 'tis nam'd:
> Now this new World was found, it was not made,
> Only discovered, lying in Time's shade. (Cavendish 2004: 121)

William Cavendish goes on to claim that Margaret's world is greater than that of Columbus, who 'Only discovered' a land, and did not invent it himself. Of Margaret's Blazing World, he says, 'But your creating Fancy, thought it fit/To make your World of Nothing, but pure Wit' (121). In one sense, he is framing Margaret's scientific ideas and fantastic adventures in the same colonialist terms as Bacon and the Royal Society. However, he is also casting Margaret's work as superior because it is generative, the product of imagination and not simple observation or conquest. The playful and satirical tone of the poem also draws attention to the critical edge of the text, which shows the horrors of colonisation and the limited value of scientific societies.

The body of Cavendish's *Blazing World* begins with a familiar element of a romance plot where an unworthy suitor abducts a distressed damsel. The 'invocation of such a plot' and 'its undoing' provides the basis for both the utopian vision and satiric twists of the story (Khanna 1994: 18). The heroic 'young Lady' is kidnapped by a foreign merchant who is 'beneath her both in birth and wealth', but there is not even a mention of a heroic man coming to her rescue (Cavendish 2004:

125). Instead, 'Heaven' disapproves of the 'theft' and 'raised such a tempest' that the boat of the kidnappers is blown toward the North Pole (125). The climate freezes all of the Lady's captors to death, but she is kept alive by 'the light of her beauty, the heat of her youth, and [the] protection of the gods' (126). The boat then moves from the North Pole to 'the Pole of another world, which joined close to it' (126). In her voyage, the 'distressed Lady' sees creatures who 'went upright as men' but who were shaped like bears, foxes, geese and many other assorted animals (126–8). Instead of being menaced by these strange new people, Cavendish represents the Lady as needing no protection. Describing natives in animalistic terms was commonplace in colonial discovery narratives, but in Cavendish's hands these bestial men actually proved to be quite gentle and technologically advanced. This account of 'multiplicity' and difference mixed elements of utopian narratives such as Godwin's *The Man in the Moone* – with its advanced and superior aliens – with the tales of savages from colonial discovery narratives (Khanna 1994: 19). Cavendish's animal men also 'anticipate the creatures from [Jonathan Swift's] *Gulliver's Travels*' as well as countless post-Darwinian SF narratives (Hutton 2003: 170–1).

Though Cavendish naturalises monarchy, the animal people of the Blazing World repeatedly prove to be superior to the men of the Lady's own world. They are described as 'good navigators' and 'experienced sea-men' who have created 'a certain engine, which would draw in a great quantity of air, and shoot forth wind with a great force' in such a way that calmed stormy seas or provided wind in the sails as needed (Cavendish 2004: 128–9). They also sail around the seas of the Blazing World with no 'guns' because 'they had no other enemies but the winds' (129). The course of science in the Blazing World is thus represented as at war with nature, just as in the Lady's world. However, there is no conflict between nations of men, and the various branches of the animal people treat each other 'with all respect and civility' (128). Cavendish's natives are not noble savages or degraded beasts: they are peaceful and technologically sophisticated beings that upset the Eurocentric pretensions of superiority common to colonial narratives of discovery.

Cavendish undermines another narrative trope when the Emperor of the Blazing World 'conceived her to be some goddess, and offered to worship her' (132). Instead of taking advantage of this colonialist conceit, the Lady 'refused, telling him … that although she came out of another world, yet was she but a mortal' (132). Rejecting the power of a goddess, the Lady is rewarded with a better alternative as the Emperor immediately makes her his wife and gives 'her an absolute power to rule and govern all that world as she pleased' (132). Now an Empress with power of state and religion, Cavendish's Lady dons an elaborate bejewelled outfit that includes a diamond shield 'to signify the defence of her dominions', and 'a spear made of white diamond, cut like the tail of a blazing star, which signified that she was ready to assault those that proved her enemies' (132-3). Instead of having her Lady become a distressed damsel in need of rescue, Cavendish has her resourceful heroine develop into an Amazon figure who eschews colonial conquest. As in her earlier writing, Cavendish here shows 'the value of the repeated metaphor ... of the warlike female for claiming equal rights for the female sex' (Venet 2003: 217).

Once her heroine is made Empress, Cavendish launches into an extensive critique of the Royal Society. The Empress begins by organising her animal-men subjects into scientific societies that discuss 'the kind of topics then being investigated by the Royal Society' (Hutton 2003: 165). Placing the Empress in the middle of such scientific debates allowed Cavendish the opportunity to experience in her imaginary world what was denied to her in the real world up to that point. Like Bacon in his utopian *New Atlantis*, Cavendish used her fictional 'narrative unashamedly as a vehicle for her own ideas' about science and its institutions (Hutton 2003: 167). However, Bacon used fiction to imagine an as yet unrealised society of men to advance his vision of science. Cavendish, on the other hand, experienced first hand what it was like to be excluded from such a utopian society. Her fictional world was the only place in which she could engage in such a utopian fellowship, though her satirical account of scientific bickering and experimentation calls into question whether she truly wanted to be involved in the regular activities of such an organisation regardless of her obvious interest in scientific subjects. The Empress finally disbands the learned societies, underscoring Cavendish's negative opinion about the value and usefulness of institutions like the Royal Society.

Cavendish did not spare herself in her satirical utopia. When the Empress decides to write 'a Cabbala', she asks spirits who would make the best scribe. The spirits warn her that 'Galileo, Gassendus, Descartes, Helmont, Hobbes, H. More, etc.' would be poor choices because they were men who were 'self-conceited' and 'would scorn to be the scribes to a woman' (Cavendish 2004: 181). Instead they endorse the Duchess of Newcastle – a comic representation of Cavendish herself – whom spirits describe as 'not one of the most learned, eloquent, witty and ingenious', but who is a better scribe due to her being a 'plain and rational writer' whose principle in writing is 'sense and reason' (181). After her soul travels to the Blazing World, the Duchess quickly becomes 'platonic lovers' with the Empress 'although they were both females' (183). This queer entanglement becomes further complicated when the Empress and Duchess commune with the soul of the Duke of Newcastle, which leads the Duchess to get temporarily jealous. The playfulness of Cavendish's self-mockery and her satirising of Plato provides yet another example of how she complicated 'gender relations' and dominant ways of thinking about intimacy and adultery (Khanna 1994: 25). It also demonstrates her self-awareness in regard to her own public image, with her critics characterising her as extravagant and vain (Sarasohn 2010: 30).

Perhaps Cavendish's most pointed critique of colonialism comes when the Duchess expresses a desire to conquer her own world so that she can become an Empress in emulation of her platonic lover. The spirits she consults warn her that 'conquerors seldom enjoy their conquest, for they being more feared than loved, most commonly come to an untimely end' (Cavendish 2004: 185). The Duchess responds that, 'I had rather die in the adventure of noble achievements, than live in obscure and sluggish security' (185). The spirits reply that 'every human creature can create an immaterial world ... within the compass of the head' (185). Instead of striving to 'conquer a gross material world' – with its endless dangers, duties and responsibilities – the Duchess resolves to 'create a world' of her own (186). This passage echoes the introductory poem, championing generative acts of imagination instead of material acts of conquest. To drive home this point, Cavendish's Empress then teams up with the Duchess to engage in a terrible campaign of violence in her original home world. This is precipitated when the Empress learns that her native country is under threat. With help from the Duchess, she conceives of and builds a submarine navy to convey her back to her old world. When confronting the enemies of her native country, the Empress again takes on the aspect of the Amazon, leading her warriors into battle dressed in an even more elaborately bejewelled outfit. This outfit includes a shield made from one fiery gem and 'a spear of one entire diamond' (211).

The knowledge produced by the dissolved scientific societies 'is appropriated for creating imperialist technologies' used to 'strengthen the Empress's imperial regime' over her home world (Walters 2014: 153). Instead of science creating a utopian dominion over nature – such as envisioned by Bacon and the Royal Society – Cavendish's satire has the Empress and the Duchess terrorise and enslave their enemies in a way that provides 'a critical view of arbitrary power' (155).

Cavendish was a remarkable woman who defied the dominant ways of thinking about gender in her society and playfully intervened into the genre system that included scientific narratives of discovery and colonial stories of adventure and conquest. *Blazing World* satirised contemporary masculine conceits about controlling nature through science and the colonial mentality of controlling people that accompanied it. Her story expounded upon the power of the imaginary world as empowering for women, providing them with realms where they may explore and rule in ways that are superior to those of men in the real world. The transgressive depictions of gender in Cavendish's work ultimately 'embraced the possibility that exceptional women can win respect and power' and 'that they are admirable' (Sarasohn 2010: 27). Cavendish's *Blazing World* began 'a continuous literary tradition' of women's writing that was transformed by Mary Shelley's *Frankenstein* and that began its full flowering in the early pulp era of the late 1920s and 1930s (Donawerth 1997: xiv).

Institutionalizing scientific masculinity and expanding discovery narratives

In spite of Cavendish's notable dissent, the Royal Society quickly became a centrally important institution that normalised and standardised scientific practices in the Anglo-American world. Perhaps the most important instance of this in the late seventeenth century came with Isaac Newton's publication of his 1672 essay 'A New Theory of Light and Colours' in the newly established scientific journal *Philosophical Transactions of the Royal Society*. Newton's famous essay highlights the growing relationship between technology and masculine discovery in the narratives of the new science at the end of the seventeenth century. In late 1671, the Royal Society witnessed a demonstration of a reflecting telescope that Newton had invented. Using the credibility and correspondence produced by the demonstration, Newton engaged in the long process of getting his 'New Theory' article considered for publication in the *Transactions*. Newton had first written about his theories and experiments on optics in his student notebooks. He expanded them in a private document, and developed them further as a series of lectures at Cambridge University (Bazerman 1988: 85-7). For the article in the *Transactions*, Newton employed a different structure than his earlier writings: he used a 'discovery narrative' where the investigator 'stumbles across an unusual fact', an approach that 'was a common one used in the early *Transactions*' (90-1). The story Newton presents of his experiments in the 'New Theory' article reorganises and reinterprets much of what he wrote in earlier versions of his work. The sequence of experiments is heavily edited, and his motivations change from the theoretical curiosity of his student notebooks to the more practical interests of an artisan working on grinding lenses. His narrative structure was literally a fiction: as Charles Bazerman concludes, 'These many discrepancies strongly suggest that Newton's discovery account was deliberately shaped for this occasion, to create the appearance of the discovery of a naturally found object, described by proper Baconian procedures' (95). The Baconian observer was implicitly masculine, and Newton's work cemented the fictional ideal of a rational and unprejudiced scientific masculinity at the centre of experimental scientific narrative.

In his 'New Theory' article, Newton's repeated evocation of his popular new telescope highlights the late seventeenth-century emphasis on practical applications of scientific work and controlling nature through the 'useful arts'. Newton's work drove home the differences between the new empirical science and the science of the ancients. As Michael Adas notes, 'The many scientific breakthroughs that culminated in Isaac Newton's experiments and writings on optics, mechanics and mathematics in the last decades of the seventeenth century left little doubt among the educated that a decisive break with the past had occurred' (1989: 71). Newton's narrative of laboratory discovery also shows the importance of the colonial discovery genre to the newly institutionalised empirical approach of early modern science. Indeed, it provides an exemplar of the new genre of the laboratory discovery narrative, a domesticated cousin of the adventurous discovery narratives of men like Columbus and Vespucci. Like those famous explorers, the practitioners of the new science were expanding the empires of men through investigation and mastery; the only difference was that their explorations occurred in the laboratory.

In the eighteenth century, the importance of scientific discovery and travel narratives grew within the European system of genres along with different iterations of scientific masculinity. The world had changed dramatically with the increasing 'exchange and interchange, circulation and transmission' of people, plants, animals, commodities, cultural practices and knowledge systems that was brought about by colonial oceanic voyages (Armitage 2002: 16). The colonial sciences of this period were not driven by some sort of insular and innocent search for the truth. As Sandra Harding shows,

> The colonists' science projects were, first and last, for maintaining Europeans and their colonial enterprises in those and other parts of the world. They were designed especially for increasing the profit Europe could extract from other lands and maintaining the forms of social control necessary to do so. (1998: 44)

The developing science, technology and medicine of Europe flourished due to the resources and knowledge Europeans gained in their voyages. Ships, navigational technologies, agriculture, plant-based medicines and countless other areas benefited from the capital poured into – and resources extracted from – colonising new lands such as the Americas (Harding 1998: 40–4). By the eighteenth century, new flora and fauna from 'voyages of discovery and the new colonies' had 'flooded Europe' and necessitated the creation of new classification systems (Schiebinger 2004b: 14).

Scientists increasingly ventured out on ships and participated in colonial voyages of discovery overseas. With few exceptions, the scientists who went on such voyages were men (Schiebinger 2004a: 235). Some eighteenth-century narratives of scientists venturing out on voyages of discovery emphasised the 'courageous heroism' of the scientists and 'made manifest a particular version of masculinity that expanded the list of desirable attributes for practitioners of science to include physical courage and fortitude as well as intellectual acumen' (Terrall 1998: 229). Long before the adventurous and heroic scientists of Jules Verne, members of the Paris Academy of Sciences were engaging in dangerous trips to Peru and Lapland to take measurements related to the shape of the earth. Part of their struggles in the new and wild frontiers came from setting up their equipment and getting reliable data. In essence, they took the laboratory to the wilderness and engaged nature in a bitter struggle for scientific truth. The stories of their struggles were widely circulated outside the academy and fuelled a new public image of scientific men of adventure (Terrall 1998: 227-37).

Taxonomy became an obsession of European scientists, and came to play an important role in the changing nature of discovery narratives of the eighteenth century. Though there were numerous classification systems produced in the 1700s, those created by Carl Linnaeus proved the most widely influential. As Linnaeus and others classified the world, they incorporated a number of existing hierarchies and ways of thinking into their systems. For example, Linnaeus easily could have chosen a different term to denote the class he came to call 'mammals' ('of the breast'): he could have chosen a term that emphasised a shared characteristic such as hair or the four-chambered heart. However, Linnaeus was caught up in ongoing debates about the importance of breastfeeding and the evils of wet nursing among European women of the middle and upper classes. By choosing to use the term mammal within his taxonomic system, he was emphasising 'how natural it was for females – both human and non-human – to suckle and rear their own children' (Schiebinger 2004b: 40-2). As such, his system of taxonomy reinforced contemporary arguments for eschewing the services of wet nurses and limiting women to the domestic sphere. The terms mammal and 'homo sapiens' ('man of wisdom') chosen by Linnaeus underscored the supposed biological inferiority of women: 'within Linnaean terminology, a female characteristic (the lactating breast) ties humans to brutes, while a traditionally male characteristic (reason) marks our separateness' from other animals (Schiebinger 2004b: 53-5).

Another such hierarchy was the perceived superiority of Europeans over peoples from other parts of the world. As European scientists tried to make sense of the new peoples encountered during voyages of discovery and colonisation, the concept of 'race' became redefined in taxonomies attempting to divide up the human species. As Nicholas Hudson demonstrates, dictionaries in French and English defined 'race' in a way that restricted the meaning 'to family lines or breeds of animals' in the late seventeenth century (1996: 247). Accounts from

European travellers before the eighteenth century also reflected this narrow understanding of race as a group of people with a common ancestry. A new meaning of the word race was used in taxonomies as far back as 1684, and this new meaning became increasingly important in the second half of the eighteenth century. Race became a general term used to separate the human species into large groups based on supposedly significant biological traits such as hair, skin colour and brain size. Differences within these races were diminished while differences between the races were exaggerated (Schiebinger 2004b: 117–20). Races were further subdivided into 'nations' or 'tribes' that shared a common language and social customs. Which term was used depended upon the place of a given group within the Eurocentric hierarchies of the day. For example, the term 'nation' began to be used only to refer to 'civilised' groups descended from European racial stock. For group distinctions made within 'savage' races, the term 'tribe' was used instead of nation. Tribes were seen as having no significant political or social organisation, and therefore constituted inferior and often nomadic peoples. By the time these terms were developed, colonisation had broken down many of the political and social structures of the peoples being colonised. This contributed to the sense that they were commonly undeveloped and 'savage' from the perspective of European commentators (Hudson 1996: 256–8).

The demand for both fictional and non-fictional travel stories exploded in the first half of the eighteenth century (Adas 1989: 69). Narratives about voyages to exotic lands reflected and reinforced the changing notions of biological difference that arose with the processes of colonisation. In addition, assessments of technology became an important part of European accounts of other peoples and their cultures. As Michael Adas shows, before the eighteenth century 'European travelers ... viewed their Christian faith, rather than their mastery of the natural world, as the key source of their distinctiveness from and superiority to non-Western peoples' (1989: 22). Though Europeans such as Bacon aspired to control nature, 'well into the eighteenth century it was not readily apparent that their level of mastery was superior to that of other civilizations, particularly those in Asia' (23). Once industrialisation began to take hold in Europe, however, this began to change. By the end of the eighteenth century, Europeans had developed a sense that their scientific achievements had 'surpassed all other civilizations' (74). In the first decades of the nineteenth century, people in

Europe – as well as in Europe's current and former colonies – 'stressed Europe's uniqueness and invariably proclaimed its superiority to even the most advanced civilized rivals' (134). Technology became associated not only with cultural superiority, but also with racial superiority in the classification systems of scientists. Male scientists and adventurers became heroic figures in narratives about the Euro-American march of civilisation: they worked in their labs and travelled the world to bring nature and 'savage' races under their mastery. This full flowering of scientific masculinity provides an important context for understanding the intervention of legendary feminist Mary Wollstonecraft Shelley.

Mary Shelley's *Frankenstein*

Mary Shelley first published her novel *Frankenstein; or, the Modern Prometheus* in 1818, and republished a significantly revised version in 1831. Since that time, the novel has become a towering presence in SF and Euro-American culture more broadly, spawning countless adaptations and reinterpretations across media. In a previous generation's debates about definitions of SF, author Brian W. Aldiss famously used Shelley's novel as the starting point for the genre. Since that time, discussions of the genre's origins invariably address Shelley's novel. Like Cavendish, Shelley represented 'the impacts of Mechanism' and modern science as 'profoundly traumatic', where 'the human subject is pierced or wounded by invasive technologies that subvert, enslave or ultimately destroy' (Luckhurst 2005: 5). For feminist SF scholars, Shelley's importance includes her extensive critique of Mechanism and modern, institutionalised science as masculinist endeavours that have dire consequences for all life. Though her novel includes a few very different scientific men, what they share is a rhetoric of colonial domination over – and penetration of – nature. Her primary scientific figures, laboratory scientist Victor Frankenstein and explorer Robert Walton, are cast as overreaching in their attempts to dominate nature, with both being taught brutally painful lessons because of their hubris. Shelley cast her critique of science in a form that has been refined by women science fictioneers ever since (Donawerth 1997: xix–xxi).

Shelley's novel immediately invokes the folly of colonial scientific conquest in an opening series of letters from Robert Walton to his sister. Walton organises a voyage to sail north from Europe and reach 'the North Pacific Ocean' because he is driven by his 'ardent curiosity' (2003: 11–12). He claims to his sister Margaret that 'you cannot contest the inestimable benefit which I shall confer on all mankind to the last generation, by discovering a passage near the pole' (11). When his ship becomes trapped in the ice, he begins to reflect critically upon his ambitions. His class position leaves him isolated and with 'no friend' among the increasingly mutinous crew (14). When Walton's men fish an exhausted Victor Frankenstein from the water, Walton finally has a friend. Walton tells Frankenstein that 'One man's life or death were but a small price to pay for the acquirement of the knowledge I sought; for the dominion I should acquire and transmit over the elemental foes of our race' (23). However, the tormented scientist chastises the explorer, lamenting 'Do you share my madness? Have you drunk also of the intoxicating draught? Hear me – let me reveal my tale, and you will dash the cup from your lips!' (23). By dramatising Walton's failures, Shelley casts a negative light on the brave, heroic version of scientific masculinity associated with voyages of discovery. Walton's colonial adventure fails due to his hubris, and Frankenstein's subsequent account of his own work links Walton's heroic version of scientific masculinity to his own more sensitive laboratory version that pays close attention to the details of nature.

Frankenstein's story to Walton shows how both heroic and sensitive versions of science are based on a common mission: the masculine conquest of nature. Recounting his conversion from hermetic to modern science at the 'university of Ingolstadt', Frankenstein tells Walton about meeting his mentor M. Waldman who he considers to be the ideal man of science. Waldman 'appeared about fifty years of age, but with an aspect expressive of the greatest benevolence' (44). In a lecture on modern chemistry, Waldman says that 'the modern masters promise very little' but 'have indeed performed miracles. They penetrate into the recesses of nature and show how she works in her hiding places' (45). These words inspire Frankenstein to 'pioneer a new way, explore unknown powers, and unfold to the world the deepest mysteries of creation' (45). Frankenstein reveals to Walton, 'I became myself capable of bestowing animation upon lifeless matter' (50). This, Frankenstein repeatedly warns Walton (and the reader), was an overstepping of natural boundaries such as life and death that is 'dangerous' and the path to 'destruction and infallible misery' (51). Shelley makes plain that part of Frankenstein's folly is his appropriation of the female domain of birth. Instead of creating a life through natural procreation, he has instead 'pursued nature to her hiding places. Who shall conceive the horrors of my secret toil, as I dabbled among the unhallowed damps of the grave, or tortured the living animal to animate the lifeless clay?' (52). The language here emphasises the Baconian tradition of seeing nature as a female to be dominated by a colonising male scientist who invades 'her hiding places'. Instead of creating life through loving union with a female nature, Frankenstein desecrates and tortures nature. In this way, Shelley dramatises the dangers of a scientific culture that excludes women and caustically comments upon hegemonic stories of male creation.

In his discussion of *Frankenstein*, John Rieder details the history of the 'divine male fabrication of human life in *Genesis* and *Paradise Lost*' where there is a 'relation between the natural and the paternal' that is represented positively and grounded in patriarchal authority (2008: 99). Drawing on the Mechanical tradition of modern science Shelley rejects this supposed authority, representing Frankenstein's science as a violation of the maternal principle of nature that usurps her life-giving essence with catastrophic consequences. Recounting his ardour for his project, Frankenstein tells of his fantasy that 'A new species would bless me as its creator and source ... No father could claim the gratitude of his child so completely as I should deserve theirs' (Shelley 2003: 52). This pursuit leads his health to fail: 'My cheek had grown pale with study, and my person had become emaciated with confinement' (52). This passage recalls the ideal of Boyle, who saw physical frailty as a sign of spiritual strength essential to his version of scientific masculinity. In Shelley's hands, physical frailty is a marker of immoral activity, immoderate scientific zeal and unnatural procreation that destroys life instead of enabling and ennobling it. Frankenstein is not just a bad mother, but also a bad father who eventually abandons his creation as soon as it comes to life. The creature's monstrous crimes are simply an outgrowth of Frankenstein's failures as a man who ignores his family and realises too late how scientific masculinity can lead to catastrophe.

Frankenstein has remained a touchstone in scientific and public sphere debates about ethics and scientific responsibility since the nineteenth century. It has also provided the template for one of the most basic sub-genres in SF: the story of mad science. This template crosses the supposed boundaries between the horror and SF genres and has been used repeatedly by both women and men. H. G. Wells famously retold Shelley's story for the age of Darwin with his second novel, *The Island of Doctor Moreau* (1896), where an unscrupulous scientist attempts to evolve 'lower' animals up to a human-like level through torturous surgical procedures. The novel dramatises how the procedures ultimately fail because it is impossible to change the evolutionary essence of a creature without horrific results. Shelley's critique of scientific masculinity's horrors provided fertile ground for women writing in the early SF pulps. Women such as Clare Winger Harris, Amelia Reynolds Long and L. Taylor Hansen portrayed scientists as threatening loners, absent-minded goofs or boring pontificators in their early fiction. Others such as Lilith Lorraine and Leslie F. Stone turned to a vision of scientific femininity that was crafted as a response to Charles Darwin's gendered vision of progress from *The Descent of Man*. This vision of scientific femininity was central to the growing branch of feminism that embraced evolutionary discourse as more modern and objective ground upon which to make the case for suffrage and social equality. The legacy of Cavendish and Shelley eventually merged with Darwinian feminism to provide a distinctive stamp to women's SF in the first half of the twentieth century.

2 Charles Darwin, Gender and the Colonial Imagination

Charles Darwin's beliefs about gender and sexuality were complex and often contradictory. Through the course of his life and career, Darwin analysed the reproductive processes and habits of scores of species. Though he found a great deal of diversity when examining the animal kingdom, Darwin's discussions of humans frequently lapsed back into broad generalisations about gender based on shaky data that seemed more driven by ideology than any kind of thorough scientific examination. He even went so far as to construct a narrative of human evolution where scientific masculinity plays a dominant role in the development of human biology and civilisation. Though his discussions of gender among humans seemed to hew closely to Victorian norms and stereotypes, Darwin's work still contained an undercurrent of heretical possibility that became a central focus of later scholars across disciplines including feminists such as Antoinette Brown Blackwell and 'sexologists' such as Havelock Ellis. The same was true of Darwin's commentaries on race and colonisation: despite using the language of racist anthropology and at times naturalising colonialism, he also engaged in pointed attacks on race as a taxonomic category and was appalled by slavery and other horrors of colonial conquest (Sharp 2007: 31–47). Darwin's complex work provided the springboard for Darwinian feminists to engage with patriarchal authority and the colonial mindset of masculinist science in a way that was not encumbered by theology and interpretations of the Bible. These feminists used Darwin's data to argue for women's control of reproduction and the development of a scientific femininity that might prove even more beneficial for progress than scientific masculinity.

Race and gender were central objects of study for natural historians and philosophers in the early nineteenth century. Despite the attempts of Euro-American scientists to define and measure race since the early eighteenth century, the concept remained very fuzzy by the time Darwin was coming of age. The meaning of the term 'race' varied from one author to the next, and was often used to mean several different things within the work of the same person (Adas 1989: 272-4; Bederman 1995: 28-9). However, there was a clear consensus among Euro-American intellectuals about many of the markers used to distinguish between the 'civilised' and 'savage' peoples of the world. One such marker was the 'arts', or what came to be known in the late nineteenth century as 'technology'. Michael Adas argues that 'scientific and technological achievements were frequently cited as gauges of racial capacity', and that such 'racist ideas in turn played a major role in late nineteenth-century debates over colonial policy' (1989: 275). Gender also came to play a significant role in the evaluation of 'savage' societies as gender ideals changed during the nineteenth century. Gail Bederman argues that the rise of the middle class in Europe and the United States was characterised by an ideal of 'manliness' that centred on the ability of men to control their impulses, work hard, 'postpone marriage until they could support a family' and save money until they could 'go into business for themselves' (1995: 12). The counterpart of this controlled manliness was the 'angel in the house', the wife and mother who was pious, fragile and economically dependent (Bederman 1995: 11; Keller 1985: 62). These assumptions about technology and gender were woven into the fabric of scientific narratives by Charles Darwin, Herbert Spencer and countless others over the course of the nineteenth century.

When Darwin went on his famous voyage aboard the HMS *Beagle* from 1831 to 1836, he took with him these assumptions and the training of a Cambridge-educated young gentleman. Indeed, it was his Cambridge connections and his gentlemanly manners that landed him a spot as a companion to Captain Robert FitzRoy on the *Beagle*. Darwin's wealthy father paid for his passage so that he could pursue natural history on the important Royal Navy surveying mission. The *Beagle* voyage that began in 1831 was the second such surveying mission undertaken with the ship. However, it was the largest undertaking yet organised by Francis Beaufort, the head of the Hydrographer's Office of the British Admiralty. Beaufort was a powerfully connected man who understood the importance of science to the military, colonial and economic interests of Britain. He pushed for training in science and mathematics for his men, as well as for the modernisation of the Royal Navy as a whole. Janet Browne argues that, 'More than anyone else [Beaufort] established the naval framework for the great age of

Victorian colonial expansion' (1995: 150). The second *Beagle* surveying mission 'would be a showpiece for the Hydrographer's Office' and Beaufort ensured that the *Beagle* 'was equipped as a mobile base for scientific instruments which were going to be used in counterpoint with other measurements taken on land' (151, 179). Beaufort selected FitzRoy as captain in large part because of his scientific training and technical expertise. When the aristocratic FitzRoy expressed the desire for a companion – a social equal who could dine with him on the ship and accompany him when on land – Beaufort put out a call for a scientifically trained 'savant' that made its way through the network of Cambridge connections. After a frenzy of letters, social calls, meetings and negotiations, Darwin became the scientific companion of FitzRoy for this crucial colonial voyage (Browne 1995: 144–61, 179–82; Clark 1984: 17–19; Desmond and Moore 1991: 101–7).

Darwin, of course, was not just going along as a companion: he was also supposed to make himself useful and engage in his own scientific studies. He shared a worktable with the ship's assistant surveyor and learned to help take measurements with the *Beagle*'s advanced scientific equipment (Browne 1995: 170–1). He also began to build his own collection of specimens to send back to the most prominent institutions in Britain, thus building his connections and reputation as a naturalist. On Royal Navy missions such as the second *Beagle* voyage, ship surgeons traditionally held the role of gathering specimens and documenting findings related to natural history. Darwin's unusual position as a gentleman and friend of the captain, however, gave him priority access to both colonial social networks and the gathering of specimens. This quickly alienated Robert McCormack, the ship's senior surgeon, who subsequently quit only four months into the trip (Browne 1995: 191–210; Desmond and Moore 1991: 123–4). Darwin was conscious of the paradigmatic differences between himself and McCormack. Darwin carried with him a copy of the first volume of Charles Lyell's *Principles of Geology*, and was greatly influenced by Lyell's emphasis on gradual change over extended periods of time (Browne 1995: 186–90; Desmond and Moore 1991: 117–18). In a letter, Darwin dismissed McCormack as 'a philosopher of rather an antient date; at St Jago by his own account he made *general* remarks during the first fortnight & collected particular facts during the last' (Browne 1995: 203). Wasting so much time in general speculation before getting to the details of empirical research was a cardinal sin for Baconian scientists, and Darwin clearly felt that

McCormack was not up to the task of such an important voyage of discovery. Darwin was happy to see McCormack go, and became the only naturalist on this Royal Navy scientific expedition that was, like others of the period, 'drawn up to fulfill complex administrative and national purposes in which geographical exploration and the rhetoric of discovery were only parts – albeit essential parts – of the developing infrastructure of empire' (181).

The scientific success of Darwin's voyage is now the stuff of legend. Upon his return, he was toasted as something of a scientific celebrity. He was elected to the Royal Society early in 1839, and later in the year published an account of his voyage that received positive reviews (Browne 1995: 407–19). Darwin was familiar with the travelogues of other scientists, and his sisters began to read through travelogues to give their brother ideas. Darwin titled his travelogue *Journal of Researches into the Geology and Natural History of the Various Countries Visited by H.M.S. Beagle*. In 1845, he revised the text (which is now referred to simply as *The Voyage of the Beagle*) to include more data that had been processed from his specimens and more analysis that was consistent with his developing ideas about evolution. He also expanded his descriptions of the natives of Tierra del Fuego and his condemnations of slavery (Browne 1995: 465–8; Desmond and Moore 1991: 327–30). Darwin's travelogue maintained the long-institutionalised form of scientific masculinity championed by Bacon, where the scientist narrator is a curious man who stumbles across natural facts and uses them to build toward larger generalisations. At the same time, Darwin's descriptions of frontier warfare, dangerous characters and treacherous landscapes added to the drama of his journey. Like the heroic scientists of the eighteenth century, Darwin was a man aboard a mobile laboratory – the *Beagle* – who put himself at risk for discovery and the progress of knowledge.

Darwin was notoriously fond of collecting insects, and early in *Voyage of the Beagle* he presents an extended discussion of the insects he observed in Brazil. In a telling passage, he describes an encounter with some ants:

> A small dark-coloured ant sometimes migrates in countless numbers. One day, at Bahia, my attention was drawn by observing many spiders, cockroaches, and other insects, and some lizards, rushing in the greatest agitation across a bare

piece of ground. A little way behind, every stalk and leaf was blackened by a small ant. The swarm having crossed the bare space, divided itself, and descended an old wall ... When the ants came to the road they changed their course, and in narrow files reascended the wall. Having placed a small stone so as to intercept one of the lines, the whole body attacked it, and then immediately retired. Shortly afterwards another body came to the charge, and again having failed to make any impression, this line of march was entirely given up. By going an inch round, the file might have avoided the stone, and this doubtless would have happened, if it had been originally there: but having been attacked, the lion-hearted little warriors scorned the idea of yielding. (Darwin 2001: 35)

Darwin's narration echoes the perspective of the Baconian experimenter found in scientific discovery narratives such as Newton's writing on optics. He represents himself as stumbling across a curiosity, and then conducting a simple experiment – blocking the path of ants with a stone – to observe their response. He then records their response and begins to formulate some generalisations about the character of the ants. Unlike the backwards McCormack, who made generalisations before detailed observations, Darwin builds his account from the ground up. At the same time, he represents this strange colonial world as filled with martial danger: even the smallest of creatures can be 'lion-hearted little warriors' who send larger creatures fleeing before their advance.

Darwin's accounts of other races are more problematic, and his developing ideas about race and gender come through clearly in his expanded 1845 account of the natives of Tierra del Fuego. Darwin makes several observations regarding what he sees as the Fuegians's poor level of technological achievement. In discussing their lack of the 'higher powers of the mind', Darwin observes that, 'Their skill in some respects may be compared to the instinct of animals; for it is not improved by experience: the canoe, their most ingenious work, poor as it is, has remained the same, as we know from Drake, for the last two hundred and fifty years' (Darwin 2001: 216). Darwin also comments that they 'have not the least idea of the power of firearms', including the danger that a gun poses to their health, in spite of repeated demonstrations (219). This is not innocent observation:

Darwin is drawing on contemporary beliefs in anthropology to size up those who he clearly sees as inferior. Their inferiority stems from their scientific failures. He compares their intellect to the 'instinct of animals' because of their inability to grasp the workings of unfamiliar technology. This comparison points toward some of his later arguments about evolution, with humanity connected to the 'lower' animals (Browne 1995: 467). Darwin also slights the ability of the Fuegians to innovate and learn through his assessment of what he sees as their highest technology, the canoe. As a result, Darwin comes to the conclusion that they are 'savages of the lowest grade' (Darwin 2001: 220). This conclusion is based on the industrial-age assumption that science and technology are the truest measure of a society and its people, and it foreshadows how later evolutionary hierarchies would cast 'savages' such as the Fuegians into a position closer to the 'lower' animals than Europeans.

Darwin develops his analysis of the Fuegians with observations about the relations between the sexes. However, his generalisations in this regard often precede his observations. In the tenth chapter of the 1845 edition of *Voyage of the Beagle*, he comments that the Fuegians,

> cannot know the feeling of having a home, and still less that of domestic affection; for the husband is to the wife a brutal master to a laborious slave. Was a more horrid deed ever perpetrated, than that witnessed on the west coast by Byron, who saw a wretched mother pick up her bleeding dying infant-boy, whom her husband had mercilessly dashed on the stones for dropping a basket of sea-eggs! (Darwin 2001: 216)

Here a generalisation about Fuegians is supported by a second-hand account. Later, Darwin notes that 'The women worked hard, whilst the men lounged about all day long, watching us. They asked for everything they saw, and stole what they could' (222). For Darwin, the Fuegian men constituted an inferior form of manhood because they did not measure up to his early Victorian ideals. Civilised men, according to the developing notion of manliness, should work hard and care for their women. For their part, women should be protected and pampered in the domestic sphere. The failure of the Fuegian men to create a 'home' that enables 'domestic affection' is for Darwin and his contemporaries a sure sign of their savage status.

Darwin concludes the chapter with an outline of what would need to happen for the Fuegians to progress up the ladder toward civilisation:

> In Tierra del Fuego, until some chief shall arise with power sufficient to secure any acquired advantage, such as the domesticated animals, it seems scarcely possible that the political state of the country can be improved. At present, even a piece of cloth given to one is torn into shreds and distributed; and no one individual becomes richer than another. On the other hand, it is difficult to understand how a chief can arise till there is property of some sort by which he might manifest his superiority and increase his power. (Darwin 2001: 229)

Darwin's conclusions demonstrate his commitment to capitalist notions of wealth accumulation and colonial notions of power. It also shows a kind of manhood he would find acceptable. A Fuegian man who showed industry (as exemplified by accumulation of wealth) would be infinitely preferable for Darwin to the loafing and stealing men he describes. Gaining material power over nature through the domestication of animals would qualify such a man to be a chief who could help his people progress, and presumably, who could provide a home for a Fuegian woman. Darwin's colonial gaze is very negative regarding the Fuegians: though he sees how they could progress under the control of a chief, he expresses serious doubts about whether they are capable of progressing.

Though Darwin expressed doubts about whether the Fuegians can progress, he supported the efforts of FitzRoy to establish a mission and help improve the Fuegians (Browne 1995: 147–9, 244–6). Darwin's version of the colonial gaze did see the Fuegians as inferior, but that did not mean he had given up on them as a people. In several sections of *Voyage of the Beagle*, Darwin attacks the colonial wars of extermination that he came across. In the fifth chapter, Darwin describes the 'banditti-like soldiers' he saw who went on 'an expedition against a tribe of Indians' in Argentina (2001: 101). Darwin draws on eyewitness accounts to describe the massacre of the Indians, which he punctuates with the observation, 'This is a dark picture; but how much more shocking is the unquestionable fact, that all women who appear above twenty years old are massacred in cold blood!' (102). Darwin's English sensibility regarding 'civilised' warfare made the murder of females seem particularly horrifying. He concluded, 'Every one here is fully convinced that this is the most just war, because it is against barbarians. Who could believe in this age that such atrocities could be committed in a Christian civilised country?' (102). Darwin makes clear in this passage that the frontier 'civilisation' of Argentina is lacking, and this is exemplified through how the colonial 'Christian' powers treat those being colonised. In particular, Darwin objected to those who were supposed to be the vanguard of civilisation mistreating and murdering women in a way he associated with the lowest savages.

Darwin had trained to be a clergyman at Christ's College, Cambridge, and had a very low opinion of those who claimed to be 'Christians' and still showed such cruelty to others. In this regard, Darwin's accounts of his voyage are in the critical tradition of Bartholomé de las Casas. A Dominican sixteenth-century priest, las Casas wrote damning accounts of the voyages of Columbus. He also gave grisly first-hand accounts of the conquest of Cuba, which emphasised the many horrible sins the Spanish committed in the process of destroying the natives (Zinn 2005: 5–7). Like las Casas, Darwin's colonial accounts in *Voyage of the Beagle* emphasise the needless slaughter of natives and the depravity of those doing the conquering. Describing the army of Argentinean General Manuel de Rosas, Darwin opined, 'I should think such a villainous, banditti-like army was never before collected together' (2001: 71). Though not particularly sympathetic to the indigenous peoples themselves, Darwin unmistakably expressed his thoughts about the ways in which colonisation should proceed. His strong support of FitzRoy's Christian mission, as well as his repeated condemnations of slavery and extermination, demonstrate that he agreed with attempts to civilise native peoples and help them progress.

Darwin's discovery narrative was very critical of certain forms of colonisation, but he never questioned the desirability of improving the natives he came across. His descriptions of the natives made plain that he viewed them through a colonial gaze: he took the measure of them by evaluating their science and technology, and he meditated on the possibilities for improving them. He was particularly hard on the men, whom he felt were responsible for their societies' lack of progress. Darwin's colonial gaze was the gendered gaze of a European scientist, and like generations of scientists before him, he assumed that science and technology was the exclusive realm of men. If a society failed in regards to science, then it was the men who were to blame for it.

That same gendered gaze led him to judge native men based on early Victorian assumptions about appropriate gender roles. This connection between science, gender and colonisation would come to play an essential role in Darwin's later writings about human evolution.

'The law of battle': sexual selection and male superiority

Darwin's account of science, gender and colonisation in his *Voyage* is just one particularly well-known example of how these ideas were taking shape in scientific discussions and discovery narratives of the nineteenth century. With the furore in the 1850s and 1860s surrounding Darwin's arguments about evolution, the already popular *Voyage* took on a new importance. Darwin had avoided the touchy topic of human evolution in his landmark 1859 book *On the Origin of Species by Means of Natural Selection, or the Preservation of Favoured Races in the Struggle for Life*. Darwin knew that his arguments about natural selection, through which favourable adaptations were preserved and disadvantageous adaptations eliminated, would be controversial enough without adding in discussions of humanity's animal origins. His arguments about natural selection proved controversial even among those who accepted his argument that species evolved over time. As Kimberly Hamlin describes the problem, 'If evolution by natural selection depended on slight variations increasing certain individuals' odds of survival, how, then, could one explain the endurance of traits such as the peacock's bright plumage and many species' large antlers?' (2010: 53). In 1871, Darwin published his two-volume work on human evolution entitled *The Descent of Man, and Selection in Relation to Sex*. In it he laid out his vision of human evolution that drew heavily from his correspondence and his experiences with other races during his voyage on the *Beagle*. He also developed a concept called sexual selection – a concept he had only mentioned in passing in *Origin of Species* – whereby mating preferences of species preserved certain physical features that otherwise would prove to be harmful to their chances of survival. By placing sexual behaviour at the centre of evolution, Darwin sparked a debate about the characteristics, behaviours and relative importance of the sexes that continues to this day.

Evolution has been proven as a fact, but details about the specific cultures and gender roles of our evolutionary ancestors have proven extremely difficult to recover. Anthropologist Sally Slocum has noted that, when reconstructing specific accounts of prehistoric societies and 'hominid evolution', scientists have to make 'speculations and inferences from a rather small amount of data' (1975: 38). To support their arguments, those who study human origins and the development of human culture have to rely on data from fossils, archaeological materials such as stone tools, 'knowledge of living nonhuman primates, and knowledge of living humans' including those living in hunter-gatherer societies (39–40). The speculative nature of working with such incomplete data has made this kind of science particularly prone to ideological influences. Problematic arguments based on assumptions about race and gender have been endemic in anthropology, archaeology, evolutionary biology and related fields throughout their histories. James Loewen has shown how some writing about history exhibits a 'rhetoric of certainty' that obscures the contingent nature of historical claims, elides the relative strength of the evidence, and ignores the possibility that ideology may be affecting the representation of the past (2007: 39–41). The same has often been true in evolutionary narratives. Rudyard Kipling's fictional 1902 children's book *Just So Stories* has frequently been cited by scholars as an example of the type of storytelling that some evolutionary thinkers lapse into when putting forward an argument. Kipling's book was full of short, fantastic origin stories such as 'How the Whale Got His Throat' and 'How the Camel Got His Hump'. Evolutionary thinkers, however, are not simply spinning Kipling-like fantasies: they are making informed speculations based on limited evidence. Unfortunately, the use of a rhetoric of certainty has led to extremely problematic narratives about the origins of human gender roles that often read like ideological fantasies about the past.

In the nineteenth century, the data on human origins that scientists had to work with were far more limited than those available today. Thomas Kuhn describes how scientific paradigms are based on work such as Darwin's that provides 'a promise of success discoverable in selected and still incomplete examples' by providing a foundation and a 'drastically restricted vision' that allows scientists 'to investigate some part of nature in a detail and depth that would otherwise be unimaginable' (1970: 23–4). The work of Darwin and his contemporaries focused later generations of scientists on the data needed to solve many of the problems with evolutionary science. Many times in *Descent of Man* Darwin draws attention to the gaps in evidence that need to be filled by research and discovery. In the first paragraph of his conclusion, Darwin maintains that, 'Many of the views which have been advanced are highly speculative, and some no doubt will prove erroneous; but I have in every case given the reasons which have led me to one view rather than to another' (629). One major problem Darwin points out is heredity: he admits that the 'laws of inheritance' are poorly understood, and acknowledges it as a significant limitation of his argument (641). This problem in evolutionary science was not adequately addressed until the early twentieth century, when Mendelian genetics was fused with Darwinist accounts of evolution and natural selection. Darwin also acknowledges the gaps in the fossil record. Though fossils of other hominids and many kinds of Palaeolithic tools had been unearthed, the paucity of such evidence in Darwin's time was noted by many of his critics. Darwin addresses this problem by declaring,

> With respect to the absence of fossil remains, serving to connect man with his ape-like progenitors, no one will lay much stress on this fact who reads Sir C. Lyell's discussion, where he shows that in all the vertebrate classes the discovery of fossil remains has been a very slow and fortuitous process. Nor should it be forgotten that those regions which are the most likely to afford remains connecting man with some extinct ape-like creature, have not as yet been searched by geologists. (163)

Since then, museums and institutes have been filled with fossils and ancient tools, evolution has become a proven scientific fact and new discoveries still make headlines on a regular basis. Darwin's frequent hedging and circumspect framing of evidence made him far more rhetorically careful than contemporaries such as Herbert Spencer.

Still, there are many points in *Descent of Man* where Darwin makes very strong assertions and uses a rhetoric of certainty. Frequently, Darwin makes claims based on the evidence provided by nineteenth-century anthropologists and sociologists. European understandings of other peoples (and of non-human primates) were still largely anecdotal by the 1870s, and did not become the focus of more systematic studies until long after Darwin's death. In the first part of the 1874 edition of *Descent of Man*, which included expanded citations of ongoing anthropological work, Darwin details the physiological, psychological and social features that connect humans to what he calls 'the lower animals' (131). In his account of human evolution, he relies on evidence from what he calls the 'savage races' and 'lower races of man' to stand in for humans in a state of nature (29, 44). Later in the work, he clarifies that, 'we are chiefly concerned with primeval times, and our only means of forming a judgment on this subject is to study the habits of existing semi-civilised and savage nations' (594). Like the anthropologists and sociologists he cites, Darwin sees human evolution as tied to racial hierarchies and technological advancement (Adas 1989: 308–9). As humans evolved, Darwin asserts that those human groups who invented and used technology better survived in greater numbers. This process of natural selection for technology had led to humans developing an upright bipedal posture, agile hands and large brains. Darwin also connects superior technology to colonisation, asserting that civilised nations had succeeded in their colonial enterprises 'mainly, though not exclusively, through their arts, which are the products of the intellect'; he goes on to associate this process with 'natural selection' (1998: 133). Thus the 'savage races' are lower down the evolutionary scale – somewhere between civilised humans and the 'lower animals' – and represent what Europeans had been like before they had progressed to their lofty position through science (Sharp 2007: 39–46). Darwin emphasises through the later sections of *Descent of Man* that the attributes associated with humanity's evolutionary advancement were more pronounced in men than in women. In effect, his narrative places long-standing beliefs about scientific masculinity as the crux of human evolutionary progress.

Darwin's argument regarding sexual selection in *Descent of Man* provided 'evolutionary biology's first universal theory of gender' (Roughgarden 2004: 164). Sexual selection also helped Darwin explain what he saw as the origins of different races (Browne 2002: 344–5). The discussion of sexual selection takes up two-thirds of the work, and begins with an overview of the 'Principles of Sexual Selection'. In the second paragraph of the overview, Darwin states that

> There are ... other sexual differences quite unconnected with the primary reproductive organs, and it is with these that we are more especially concerned – such as the greater size, strength, and pugnacity of the male, his weapons of offence or means of defence against rivals, his gaudy coloring and

various ornaments, his power of song, and other such characters. (1998: 215)

This image of the active, virile male dominates Darwin's image of sexual selection. According to Darwin, males developed these characteristics 'not from being better fitted to survive in the struggle for existence, but from having gained an advantage over other males' (217). It is therefore the drive to win a mate from rivals that Darwin cites as the cause for such pronounced sex differences. He also claims that 'it is, with rare exceptions, the male which has been the more modified' by sexual selection (229). Darwin therefore sees males as evolving due to the competitions of sexual selection while the females remain relatively plain and undeveloped.

Females, for their part, are seen by Darwin as having some power in regard to mate choice. He asserts that,

> It is certain that amongst almost all animals there is a struggle between the males for the possession of the female. This fact is so notorious that it would be superfluous to give instances. Hence the females have the opportunity of selecting one out of several males, on the supposition that their mental capacity suffices for the exertion of choice. (1998: 219)

Though definitely not as important as the male role, Darwin's assertion that females had a power of choice proved a point of great importance for later feminists. So did Darwin's characterisation of males. In the struggle to reproduce, Darwin avers that the active 'males of almost all animals' have evolved 'stronger passions than the females' (229). Contrary to these passionate males, Darwin posits that, 'The female ... with the rarest exceptions, is less eager than the male' (230). He goes on to say that 'she is coy, and may often be seen endeavouring for a long time to escape from the male' (230). Despite this coy behaviour, Darwin adds that 'the female, though comparatively passive, generally exerts some choice and accepts one male in preference to others' (230). In the seventeenth century, many scientists saw women as deceptive, Eve-like temptresses who were the enemies of chaste scientific work. The witch, a female figure that was characterised by 'unbridled sexuality', became the focus of acrimony for many seventeenth-century scientists (Keller 1985: 59). Darwin's representation of females as coy, passive and chaste demonstrates the marked changes that had occurred in ideologies of gender by the Victorian period. At the same time, his image of males as amorous, active and violent became another starting point for feminist critiques of male social and political hegemony.

After stating the biological principles he sees at work in the sexes, Darwin makes his way through the animal kingdom cataloguing examples that prove his point. Beginning with 'the lower classes of the animal kingdom', Darwin argues his way up to the pinnacle of evolution: humanity (1998: 268). Along the way, he discusses molluscs, crustaceans, spiders, insects, fishes, amphibians, reptiles, birds and mammals. Among most non-human animals, Darwin asserts that males have become more beautiful and more virile in the contest for coy females. This dual male struggle to battle other males and entice females led to characteristics of both the 'warrior' and the 'dandy'. However, Darwin assures the reader that even though 'The peacock with his long train appears more like a dandy than a warrior ... he sometimes engages in fierce contests' (375). For humans, this situation is more complicated. In the opening to his section about mammals, Darwin states that, 'With mammals the male appears to win the female much more through the law of battle than through the display of his charms' (518). In effect, he states that male mammals are much more warrior than dandy. Despite the 'fact' that savage 'men are more ornamented than the women' in most places of the world, Darwin extends this argument to the highest of mammals (598). Drawing on anthropological accounts, Darwin argues that, 'With savages, for instance, the Australians, the women are the constant cause of war both between members of the same tribe and between distinct tribes' (581). Where females in the lower animals have the power to choose a mate, for humans this situation has changed:

> Man is more powerful in body and mind than woman, and in the savage state he keeps her in a far more abject state of bondage than does the male of any other animal; therefore it is not surprising that he should have gained the power of selection. (619)

With men in complete control of sexual selection, Darwin posits that 'women have long been selected for beauty' and 'thus have become more beautiful, according to general opinion, than men' (619). As such, humans seem to be the exception in Darwin's picture of the animal kingdom.

Darwin connects this argument about sexual selection to his earlier points about technology and human evolution, making it abundantly clear that human males were the ones naturally selected to invent and use technology. He links this male ability to their supposedly superior manual dexterity and brains:

> Man is more courageous, pugnacious and energetic than woman, and has a more inventive genius. His brain is absolutely larger, but whether or not proportionately to his larger body, has not, I believe, been fully ascertained ... The female, however, ultimately assumes certain distinctive characters, and in the formation of her skull, is said to be intermediate between the child and the man. (1998: 577)

Though showing some scepticism about the quantifiable nature of intelligence in relation to brain size, Darwin states plainly his belief that males have superior brains. By commenting on the formation of the skull Darwin is evoking the science of craniometry, where skull measurements are used to assess intelligence and abilities (Gould 1996: 62–104). In this case, Darwin uses craniometry to add strength to his depiction of human females as less evolved intellectually than males. A few pages later, Darwin breaks down some of the unique mental abilities of women:

> It is generally admitted that with woman the powers of intuition, of rapid perception, and perhaps of imitation are more strongly marked than in man; but some, at least, of these faculties are characteristic of the lower races, and therefore of a past and lower state of civilization. (1998: 584)

In this one sentence, Darwin dismisses both white women and the non-white races as being evolutionarily backward.

In case his point is still unclear, Darwin goes on to add, 'The chief distinction in the intellectual powers of the two sexes is shown by man's attaining to a higher eminence, in whatever he takes up, than can woman – whether requiring deep thought, reason, or imagination, or merely the use of the senses and hands' (584). Where women and 'the lower races' are intuitive, white men use reason; where women and 'the lower races' are 'imitative', white men are imaginative and inventive. By stating that white men have superior intellectual skills and manual dexterity, Darwin marks them as the biologically superior toolmakers. He thereby excludes all women and non-white men from having much relevance to the 'man the toolmaker' narrative of human evolution. For Darwin, women are physically unsuited for handling technology well and intellectually incapable of excelling in scientific endeavours (Deutscher 2004: 44–5; Hubbard 2002: 160–1). Darwin also reinforces long-standing assumptions about scientific masculinity, because it is only white males – and not men such as those from Tierra del Fuego – who have the right qualities for making contributions to science. Because of these processes of sexual selection, Darwin concludes, 'man has ultimately become superior to woman' (1998: 585). This is not the language of cautious, circumspect speculation: this is the rhetoric of certainty, with Darwin taking it as an obvious fact that men are more important than women in the material progress of civilisation. Because of their biology, Darwin asserts that women are nurturing, beautiful prizes to be fought over by the men for reproductive purposes. This means that (white) men are the natural scientists and soldiers, not women (Hubbard 2002: 158–61; Roughgarden 2004: 164–72).

In fleshing out his image of the innate characteristics of the human sexes, Darwin did not go much farther than the standard 'Victorian stereotype of the active male and the passive female' (Hubbard 2002: 161). His evolutionary narrative projects these stereotypes into the past and attempts to muster biological reasons for what he sees as the current realities of sexual difference. Darwin states that 'in the case of mankind', men had

> to defend their females, as well as their young, from enemies of all kinds, and to hunt for their joint subsistence. But to avoid enemies or to attack them with success, to capture wild animals, and to fashion weapons, requires the aid of the higher mental faculties, namely, observation, reason, invention, or imagination. These various faculties will thus have been continually put to the test and selected during manhood; they will, moreover, have been strengthened by use during this same period of life. Consequently in accordance with the principle

often alluded to, we might expect that they would at least tend to be transmitted chiefly to the male offspring at the corresponding period of manhood. (1998: 584-5)

Darwin saw technology as important for males because they could use it for violent purposes (Hubbard 2002: 160). The violent struggle of the males had changed their bodies over time, Darwin argues, with the brain and hand becoming more important as technology replaced biological weapons such as fangs (1998: 54). Darwin's notion of heredity was woolly by modern standards, and he believed that males somehow inherited the traits of their fathers more than those of their mothers. As such, he believed that men were by nature violent, selfish toolmakers who passed these traits on to their sons (Deutscher 2004: 44). This image of manhood was consistent with generations of Euro-American scientists who linked science, gender and colonialism in their accounts of discovery and the natural world. Human evolution was the domain of superior men whose scientific discoveries and technological mastery made them the masters of nature, women and other races. In effect, Darwin's argument helped naturalise representations of science, gender and colonialism that had been circulating through Euro-American systems of genres since before the Scientific Revolution.

Motherhood and evolutionary essentialism

Amy Kaplan argues that, in the writing of several nineteenth-century women authors, domesticity 'related to the imperial project of civilizing' and

> the conditions of domesticity often become markers that distinguish civilization from savagery. Through the process of domestication, the home contains within itself those wild or foreign elements that must be tamed; domesticity not only monitors the borders between the civilized and the savage but also regulates traces of the savage within itself. (1998: 582)

This colonial vision of domesticity also holds true with the writing of nineteenth-century evolutionary thinkers. The status of women played an important role in how travellers, anthropologists and naturalists such as Darwin evaluated foreign cultures while defining the 'natural' role of women in their own. In his *Voyage of the Beagle*, Darwin represented the Fuegians as inferior in part because of the failure of their men to establish a 'home' for their women. In *Descent of Man* he repeats his assessment of the Fuegians and how the savage man 'treats his wives like slaves' (1998: 643). Women working outside a recognisable home in a hunter-gatherer society constituted slavery for Darwin, who asserted that men protecting their women in a safe, domestic space was an essential requirement for civilisation.

Like Linnaeus and countless other scientists, Darwin also defined women in terms of their biological role in reproduction. In his vision of sexual selection, women were the catalyst for fighting between the men. More significantly, Darwin connects the importance of women to their role as mothers. Where man is selfish and violent, woman has 'greater tenderness and less selfishness' and 'owing to her maternal instincts, displays these qualities towards her infants in an eminent degree; therefore it is likely that she would often extend them towards her fellow-creatures' (1998: 583). In an earlier chapter, Darwin juxtaposed the cruelty of savages with 'the instinct of sympathy' that permeated the 'civilised nations' (138–9). For Darwin, sympathy 'is the noblest part of our nature', and as the 'sympathies' of humans 'became more tender and widely diffused ... so would the standard of ... morality rise higher and higher' (129, 139). Therefore, the key to human morality is a trait that he connects to motherhood. With women the ones who extend sympathy, it is through them that morality progresses and the savage cruelties of human nature are overcome. In this sense, domestic space and the 'angel in the house' are central to Darwin's vision of civilisation, and the white mother is the agent and carrier of moral progress.

Evolutionary ideas became more widely distributed in the United States with the publication of *Popular Science Monthly* beginning in 1872. The journal was created as a means for distributing the ideas of Herbert Spencer and his views on sociology, but numerous commentators from various fields contributed their insights as well. Edited by E. L. Youmans, *Popular Science Monthly* 'attracted 11,000 readers in its first year' and helped fuel sales of Spencer's books (Bannister 1979: 59). Despite the wide audience Spencer and his followers had, his ideas were received with a great deal of scepticism. As Robert C. Bannister argues, 'most of Spencer's disciples operated outside the colleges or the churches' and 'No leading Spencerians were scientists' (61). Despite their lack of institutional or scientific authority, the ideas of Spencer and his followers helped shape broader cultural conversations about evolution and gender. Spencer agreed with some feminist ideas as late as the 1850s, but by the time *Popular Science Monthly* began publication his ideas about the sexes had changed radically (Paxton 1991: 15).

In the November 1873 issue, Spencer published an essay entitled 'Psychology of the Sexes' that made plain his views about the 'natural' status of women in society. Like Linnaeus, Darwin and countless others, Spencer anchored female identity in the biological and social imperatives he associated with reproduction and motherhood. However, Spencer's account of evolution was even less flattering to women than Darwin's. In the second paragraph of the essay, Spencer states that, 'Just as certainly as [men and women] have physical differences which are related to the respective parts they play in the maintenance of the race, so certainly have they psychical differences, similarly related to their respective shares in the rearing and protection of offspring' (31). In a lengthy footnote at the end of the paragraph, Spencer criticises approaches that are more flattering to women. In particular, he argues for the importance of understanding the 'normal mental power' of the sexes instead of focusing on exceptional individuals who deviate from the norm (31, n. 1). Spencer gives an 'extreme case' to make his point:

> the mammae of men will, under special excitation, yield milk ... But this ability to yield milk, which, when exercised, must be at the cost of masculine strength, we do not count among masculine attributes. Similarly, under special discipline, the feminine intellect will yield products higher than the intellects of most men can yield. But we are not to count this as truly feminine if it entails decreased fulfillment of maternal functions. Only that mental energy is normally feminine which can coexist with the production and nursing of the due number of healthy children. (31, n. 1)

Spencer represents biological energy as a zero-sum game where energy needed to gestate and nurse children detracts from mental abilities. In this sense, being a woman means having reduced energy for intellectual pursuits compared to men so that her body can carry out what Spencer sees as her biological purpose: motherhood. This makes it clear what Spencer thinks about women's education, which risks spending women's energies on abnormal intellectual pursuits instead of 'normally feminine' mental pursuits that are consistent with motherhood.

Spencer further elaborates upon why he believes men are more developed than women: 'in woman, an arrest of individual development takes place while there is yet a considerable margin of nutrition: otherwise there could be no offspring. Hence the fact that girls come earlier to maturity than boys' (1874: 32). The 'fact' that men continue to develop means that they manifest 'the latest products of human evolution – the power of abstract reasoning and that most abstract of emotions, the sentiment of justice – the sentiment which regulates conduct irrespective of personal attachments and the likes or dislikes felt for individuals' (32). This means that for Spencer, as for Darwin, women are less capable of engaging in pursuits associated with science and invention such as abstract thought and objectivity. Instead, Spencer argues that women developed other 'mental traits' that emerged from 'early barbaric times' when women had to appease 'the men of the conquering races which gave origin to the civilized races' (33). To deal with such 'merciless savages' as these men, Spencer asserts that women evolved 'the ability to please, and the concomitant love of approbation' (33). In other words, women learned how to be happy doormats in the service of their savage husbands in order to survive. Spencer here appeals to a generic narrative of colonial anthropology without even the benefit of a specific example. When Darwin made such statements he often qualified them or pointed out the limitations of existing evidence that supported his conclusions. Spencer's writing exhibits no such qualms, stating all things with a strong rhetoric of certainty and marshalling few (if any) concrete examples from nature to support his large generalisations.

This is one reason why some feminists gravitated toward Darwin: his appeals to specific examples from nature opened the door for counter-examples that could dispute his conclusions about gender and the relative superiority of men. Also, despite asserting that men have greater 'intellectual powers' than women, Darwin made an argument in favour of women's education. In a chapter of *Descent of Man* entitled 'Mental Powers of Man and Woman', Darwin speculated that,

In order that woman should reach the same standard as man, she ought, when nearly adult ... to have her reason and imagination exercised to the highest point ... All women, however, could not be thus raised, unless during many generations those who excelled in the above robust virtues were married, and produced offspring in larger numbers than other women (1998: 586)

Darwin still saw the value of women's education through the lens of maternity – and his Lamarckian understanding of heredity – by asserting that women's intellectual gains would only be beneficial if they passed them on to many children. However, Darwin saw education as something that could be additive for women instead of casting it as something that would drain women of reproductive power. This would prove important for feminists such as Antoinette Brown Blackwell.

The work of men such as Darwin and Spencer helped entrench an evolutionary essentialism in Euro-American culture. This evolutionary essentialism posited that each of us have within us the legacy our animal ancestry; despite the fact that species may evolve over time, individuals could not change their own evolutionary essence (Kaplan and Rogers 2003: 28). The restraints of civilisation – as epitomised by domesticity – help keep the animal impulses of humans in check, but at root both Darwin and Spencer saw every man as harbouring a selfish, violent nature. Darwin and Spencer thought women were beautiful spurs to male violence as well as selfless, intellectually inferior balms for the virile drive toward conflict at different points in their biological development. This evolutionary essentialism still contained the hope that things could change for the better, and Darwinian feminists in particular saw the possibilities for progress within the framework of sexual selection established by Darwin's *Descent of Man*. While more conservative feminists gravitated toward Spencer's model of submissive domesticity and automatic social development, Darwinian feminists appealed directly to nature and Darwin's model of argumentation in making their claims for emancipation.

Antoinette Brown Blackwell's *The Sexes throughout Nature*

The first feminist to directly engage in a published critique of Darwin's *Descent of Man* was Antoinette Brown Blackwell, who published her essay 'Sex and Evolution' as a part of her book of essays *The Sexes throughout Nature* in 1875. The book also included one essay published previously in *Popular Science Monthly* and others that were published in *Woman's Journal*, the 'official paper of the American Woman Suffrage Association (AWSA)' (Hamlin 2014: 37). However, 'Sex and Evolution' was an original piece that took up more than half of the book, and that established a number of critiques of evolutionary science that have remained consistent among feminists ever since. Though excluded from membership in scientific institutions, educated women had been active in writing, publishing and reading 'popular' scientific articles since the eighteenth century (Gates and Shteir 1997: 6–10). For much of the nineteenth century, educated women in Britain and North America 'functioned not as the groundbreakers but as educators, carefully explaining new views of the physical and natural world to women, children, and the working classes' (10). Blackwell and fellow feminist Eliza Burt Gamble, however, charted a departure from this as they actively worked to construct a scientific femininity that did not simply repeat the work of scientific men, but rather added to it in ways that provided a more complete view of nature. These women became part of a larger movement to provide formal scientific training to women and promote the participation of women in scientific professions.

Born in 1825, Antoinette Brown grew up in Henrietta, New York, where she had a relatively privileged and religious upbringing, graduating from the Monroe County Academy at the age of fifteen (Cazden 1983: 12–13). She worked as a teacher after graduation, and eventually followed her older brother William to study in Ohio at Oberlin Collegiate Institute, one of the few places that allowed women to study alongside men. Oberlin's president, Charles Grandison Finney, 'had been the Brown family's spiritual leader' and the liberal, anti-slavery activism of the campus suited the 20-year-old Antoinette perfectly (16). Though she did not take the more science-oriented course of classical training, she did study natural philosophy as a part of her plan to pursue theology. Oberlin was undoubtedly progressive for its day, but that did not extend to condoning women speaking in public or in classes, and it did not ordain women as ministers (25–6). Antoinette still pursued theology with the intention of preaching, and after completing her studies at Oberlin began a career as a public speaker on the lecture circuit with the hopes of finding a church to ordain her as its minister (58–9). In 1853, Antoinette Brown accepted an invitation to lead a church in South

Butler, New York, where she was ordained as a minister in the orthodox Congregational Church (74-8). However, she soon found herself losing faith in organised religious orthodoxy, leaving her ministry and returning to work as a lecturer and writer on issues such as poverty, abolition and women's rights (Cazden 1983: 92-7; Hamlin 2014: 61). In early 1856, she married fellow progressive Samuel Blackwell, a member of a well-known abolitionist family who had five feminist sisters, two of whom were medical doctors (Cazden 1983: 98-9, 108-9). She had seven children with Samuel, but continued to speak and write, her work eventually turning toward science to help study questions that her religious training had not adequately answered (Hamlin 2014: 102).

Blackwell was a pioneering woman who refused to accept the idea that women could not be ministers, and she showed similar disdain in *The Sexes throughout Nature* toward the idea that women could not be scientists. Blackwell begins 'Sex and Evolution' by singling out the arguments of both Darwin and Spencer as the products of a suspect form of scientific masculinity. She shows great faith in objectivity and the scientific method, claiming that 'the special class of feminine instincts and tendencies' she proposes is 'a question of pure quantity' that 'can be experimentally decided, and settled by rigidly mathematical tests' (Blackwell 1875: 11). However, she points out that Darwin and Spencer fail to address 'the dogma of male superiority', and instead perpetuate 'the male standpoint' of older scientific works (14, 17). In place of this male standpoint, Blackwell claims that 'there is no alternative' but for women to engage in science, because 'Only a woman can approach the subject from a feminine standpoint' (22). In this way, Blackwell directly engages the performativity of scientific discourse, calling into question the observer of scientific 'facts' and emphasising how 'observations are embedded in the dynamic relationship between observer and environment' (Bruni 2014: 14). This draws attention to the reality that for the reader, this is not first-order observation, which is 'an observer observing objects', but actually second-order observation, 'where observers watch other observers' (2014: 5). In other words, Blackwell values objectivity, but does not trust the linguistic games of scientific discourse where the scientist pretends to be some sort of transparent observer presenting facts for readers in a way that allows readers to see objects for themselves. Instead, the 'male standpoint' of the scientist is a coloured lens, and Blackwell underscores how reading the work of men such as Darwin and Spencer is actually watching an observer study an object, not directly studying the object itself. Blackwell rejects the bias implicit in male scientists attempting to discuss women when they have no direct experience of what it is to be a woman, and instead 'reinterpreted' the evidence presented by Darwin and others 'from a female standpoint, often offering her own experiences in support of her claims' (Hamlin 2014: 104). In effect, Blackwell challenges scientific masculinity as limited, and performs a complementary form of scientific femininity to provide a more complete picture of nature and of women.

Blackwell's study of scientific writings and the animal world 'did not convince her that females were equal to males', but rather 'that the sexes in all species were different but complementary to each other' (Hamlin 2014: 105). Blackwell points to the 'division of labor' between sexes and the importance of females providing '*direct nurture*' and '*direct nutrition*' to their young through functions such as lactation, while males of some species provide 'largely for their *indirect nurture*' through bringing home food for the family (Blackwell 1875: 30–1, 41). The expenditure of energy for these functions, she explains, does not rob females of their other faculties. Referring explicitly to the male focus in Darwin's account of sexual selection, Blackwell points to the 'equally important' characteristics of 'mothers' who 'act with the utmost wisdom and good faith and with a beautiful instinctive love towards a posterity which they are directly never to caress or nurture' (61). Blackwell praises the 'parental love' of females as opposed to the mere 'sexual love' of males, pointing out that females in many species are the ones who build homes and organise societies (62–3). The 'sexual fervor' of males therefore is simply a somewhat limited complement to this more constructive female characteristic (65). For Blackwell, 'it was the nutritive function provided by mothers that signalled evolutionary advancement' (Hamlin 2014: 106). Motherhood was the core of Blackwell's account of progress and civilisation, and in this way she inverted the Darwinist argument of 'man the toolmaker' that saw men as the prime movers of evolutionary progress.

Blackwell did engage in a type of evolutionary essentialism that accepted the premise that the minds of men and women are basically different (Murphy 2006: 12). In the case of humans, however, Blackwell rejected Spencer's assertion that female energy directed toward intellectual pursuits was unnatural. Blackwell argues that, feminine functions find their place in the system, co-ordinated with all the others ... Periodicity of function, maternity, lactation, all being organically provided for, each in relation to the other, neither should cause the least disturbance to the health; neither should subtract anything from the general functions of nutrition, and all should add, as all other balanced activity does, to a larger vigor both of mind and body. The legitimate use of every faculty is strengthening, not exhausting. (1875: 110-11)

For Blackwell, biological energy is not a zero-sum game as it was for Spencer. Like Darwin, she calls for women to exercise 'all their faculties', including their intellects, to strengthen themselves and their children (115). Blackwell then challenges that, 'It may be found, process for process, in detail and in totality, that the average woman is equal to the average man. By all means let the sexes be studied mathematically' (111). Blackwell thus reinforces her faith in the objectivity of science, and calls into question Spencer's purely theoretical claims about the energies and natural capabilities of the sexes.

Blackwell also rejects Spencer's assertion that women are incapable of objectivity because of their less developed biology and relative emotional immaturity. Male reasoning is an 'indirect method of acquiring truth', she asserts, whereas Darwin's account of woman's essence provides the key to understanding how she is more empirically minded (1875: 120). Darwin claimed that women are more intuitive than men, and Blackwell extends this by saying that the 'rapid intuitions with which women are credited, are simply direct perceptions' (119). This means that women make better and faster empiricists than men, so Blackwell concludes 'it must become evident that [women] would bring new modes of force and fresh methods of inquiry into every department of research' (121). Blackwell later elaborates that, 'woman, as much or more so than any other being, man not excepted, is able to put aside personal feeling and personal interest, to follow the lead of impersonal judgment' (132). Alluding to Darwin's assertion of essential male selfishness, Blackwell points out that 'woman is not constitutionally self-centred in thought or feeling' due to her maternal essence: woman's 'instincts impel her to self-forgetfulness in thinking and acting for her children, and inherited habit has developed and extended the tendency to whatever person or subject ... occupies her thoughts'

(133). In this way, Blackwell uses Darwin's essentialist gender formula to refute Spencer and claim for women the mantle of the sex that is by nature 'objective in thought' (133). By doing so, she implies that scientific femininity could be more powerful and produce greater results than scientific masculinity.

Near the end of her essay, Blackwell points out how unscientific the assertions of Darwin and Spencer are regarding women because they do not take into account the social impediments to women's intellectual and artistic achievements. She asserts that male superiority cannot be proven unless woman 'is allowed an equally untrammelled opportunity to test her own strength' and 'a running comparison' made between men and women (1875: 135). This call for equality of opportunity was a common feminist refrain, but Blackwell made it in the context of a compelling evolutionary argument supported by many biological examples. Domestic labour was one significant impediment that Blackwell took time to comment upon, noting that 'it seems to me that fathers are equitably bound to contribute indirect sustenance to offspring in the shape of good edible food for the mother. To this we might add ready-made clothing and fires lighted on cold winter mornings!' (114). In her own life, she had experienced the benefits of this as her husband Samuel 'shared in the work of the home willingly and without argument' (Cazden 1983: 162). The redistribution of domestic labour was a major feminist talking point for the next several decades, as it was directly connected with the ability of women to pursue lives and careers that went beyond the home.

Blackwell's performance of scientific femininity provided encouragement for other women to do the same, weighing their own experiences as women against the images of sex and nature presented by scientific men. As Hamlin observes, 'Blackwell pioneered the practice of looking to the animal kingdom for alternative examples of domestic arrangements, a strategy that proved inspirational and rhetorically powerful for nineteenth century feminists' (2014: 105). Chapters three through five demonstrate that this practice also proved indispensable for women writing SF in the early twentieth century as they appealed to nature and imagined alternative pathways for the evolution of sexual relationships and domestic arrangements. Blackwell's desire for women to take up scientific pursuits was shared by 'many of the nation's leading women', who helped build up scientific facilities and curricula at women's colleges such as Smith in the late nineteenth century (Hamlin 2014: 63). The development of women's science education had a direct influence on early SF, with female authors such as Clare Winger Harris, Lilith Lorraine and L. Taylor Hansen benefitting from formal training and using it to spin scientifically engaged futures that furthered the arguments of Darwinian feminism.

Eliza Burt Gamble's *The Evolution of Woman*

Nearly twenty years after Blackwell published *The Sexes throughout Nature*, feminist Eliza Burt Gamble upped the ante with her widely read *The Evolution of Woman: An Inquiry into the Dogma of her Inferiority to Man*. In this book, which she revised in 1916 as *The Sexes in Science and History*, Gamble further developed evolutionary arguments about motherhood as the cornerstone of civilisation. The importance of matriarchal society had been popularised earlier in the century by Swiss scholar Johann Jakob Bachofen's *Das Mutterrecht* (1861). Drawing on classical sources, Bachofen developed an account of matriarchy that enshrined 'mother love' as the origin of the first social progress. Early society, he argued, had been a savage, male-dominated affair where women were basically slaves. This eventually gave way to matriarchy when women demanded monogamy, and society grew around this mother love to reach a new level of civilisation. This eventually gave way to a new kind of patriarchy where men dominated marriage and property, something that Bachofen saw as natural (Eller 2011: 41–7). Many anthropologists adopted Bachofen's stages of social development along with his reading of the last patriarchal stage as progress. Socialists and feminists, however, reinterpreted the last stage of his history as 'a myth of regress rather than a myth of progress': they saw matriarchy in terms of socialist collaboration and patriarchy in terms of capitalist selfishness (2011: 101). Gamble was drawn to Darwin's idea of sexual selection – which emphasised the brutality and enslavement of women by men – as a scientific and valuable evaluation of sex that confirmed her socialist beliefs about early matriarchal society and the violent selfishness of modern capitalism.

Instead of following Blackwell's theory of equal complementarity, Gamble stressed the superiority of women over men (Jann 1997: 150). Gamble argued throughout her adult life that 'culture and economic power constructed female weakness', not biology, and she rejected the belief that 'modern property relations' were the 'highest development of human civilization' (149). Gamble was clear on the importance of socialism and strong motherhood because of her own life: she was brought up by 'a widowed mother' and was 'orphaned as a teenager' (Hamlin 2014: 129). Born in Concord, Michigan, Eliza Burt 'taught in various Concord public schools and was a superintendent in East Saginaw' (Jann 1997: 149). She married James Gamble in 1867 and had three children, one of whom died at a young age. With her husband's support, Gamble began to develop her socialist feminist ideas through extensive research and writing. She both organised and presented at meetings promoting women's suffrage in the Midwest, and beginning in 1885 spent a year reading at the Library of Congress. It was during her year in Washington DC that she discovered the work of Darwin, which provided the key she was seeking for her work (Hamlin 2014: 128–9, 137–9; Jann 1997: 149).

In the preface to *The Evolution of Woman*, Gamble emphasises the centrality of Darwin to her work when she states that it was 'after a careful reading of *The Descent of Man*, by Mr. Darwin, that I first became impressed with the belief that the theory of evolution, as enunciated by scientists, furnishes much evidence going to show that the female among all orders of life, man included, represents a higher stage of development than the male' (Gamble 1894: v–vi). In the body of her argument, Gamble discusses how the traits ascribed to women by Darwin made them uniquely qualified to control society. Gamble notes that in Darwin's argument, 'sympathy constitutes the foundation-stone of the social instincts' (57). The origin of sympathy is with women, who have 'acquired characters tending toward the welfare of society, or of individuals outside of the self', whereas men have acquired 'characters looking only toward selfish gratification' (57). Gamble concludes that, 'Although Mr. Darwin does not admit it, from his reasoning it is plain that the maternal instinct is the root whence sympathy has sprung, and that it is the source whence the cohesive quality in the tribe originated' (61). For Gamble, both progress and civilisation were carried within women, and it was a simple matter to connect the dots that Darwin himself had not. To oppress women and keep them from their full potential was to stunt progress and the evolution of the species.

Gamble went on to catalogue not only the superior traits of women, but also many of the traits that in her view made men biologically inferior. Gamble quotes Darwin's claim that women across races are less hairy then men, and then points out Darwin's analysis that 'hairiness denotes a low stage of development' (1894: 41). More significant, however, is the pervasive evil of male sexual desire. Echoing Blackwell, Gamble posits that male evolution, which was driven by their attempts to attract and control females, caused them to develop physical features that have 'been acquired at the great expense of vital force' (25). The greatest sign of male inferiority – and for Gamble, the most regressive force in human history – is the 'abnormal development of the reproductive energies' of men, which has fuelled their drive to capture women for wives and thereby 'drain' the 'physical energies' of women (68). One essential female trait Gamble sees across species is that for them 'pairing is not only a matter of indifference', but in general 'courtship is actually distasteful to them' (16). This representation of the sensual male and the chaste female was common in the social purity crusades across women's movements in the late nineteenth century, and Gamble saw the New Thought movement as an ally and audience in her push to contain male immorality and emancipate women (Satter 1999: 112–14). She also painted a clear picture of males as biologically wasteful and selfish, whereas her account of females showed how they use 'vital force for more useful functions' (Hamlin 2014: 135). The supposed inferiority of women in Darwin's work was actually his misreading of his own data, and Gamble expanded upon Darwin's failure to account for social factors associated with patriarchy that drain and enfeeble women in their attempts to compete with men in intellectual and professional pursuits.

Gamble again followed Blackwell (and countless other feminists) in *The Evolution of Woman* by singling out women's clothing as a particularly illustrative example of how social forces perpetuate the enslavement of women and damage their health. Women's clothing, she reasons, must be understood in terms of Darwin's sexual selection argument, and had developed along lines that serve patriarchy:

> The processes of sexual selection, which, so long as the female was the controlling agency in courtship, worked on the male, have in these later ages been reversed. For the reason that the female of the human species has so long been under subjection to the male, the styles of female dress and adornment which have been adopted, and which are still in vogue are largely the result of masculine taste. (1894: 69–70)

Men controlling sexual selection, she reasons, has compounded the humiliations women must undergo in order to survive. This has led to 'pernicious and health-destroying styles', and negative consequences for 'women who have been sufficiently courageous to attempt to free their ankles from the cumbersome skirts so detrimental to health and so destructive to the free use of the legs' (71). Returning control of sexual selection to women is the only way Gamble sees that women can be freed from the oppressiveness of male tastes and sexual desires, and the crux of her argument is that this would lead to both social and biological progress.

Eugenics had a strong appeal for many reform-minded feminists, and Gamble was no exception, but there were a wide variety of opinions between commentators over how eugenics should function. Francis Galton, Darwin's cousin, came up with the term 'eugenics' in 1883 to describe how humans could gain 'control over' something that was a 'heretofore untamable process' (Rosen 2004: 5). His explanation of positive eugenics sought to encourage the most evolutionarily fit people to breed with one another, whereas his formulation of negative eugenics urged preventing those who were unfit from reproducing at all. Feminists saw that controlling reproduction could have a number of tangible benefits for women. Pregnancy and childbirth were dangerous, with 'seven deaths per one thousand live births' in the United States at the end of nineteenth century (Hamlin 2014: 137). Controlling reproduction would also allow women to plan their families more carefully and avoid becoming financially destitute baby factories. Eugenics provided a scientifically compelling argument for controlling reproduction that many feminists embraced as a means to limit women's exposure to health risks and expand their economic opportunities. Many 'feminist socialists' believed in 'empowering individual women to make the best choices for themselves' in regards to how, when and with whom they should (and should not) have children, where others looked more to state-level control of reproduction (142). Gamble was a Darwinian feminist who fell clearly into the former camp, arguing for most of her adult life that women's control of marriage and reproduction offered benefits for society and for individual women.

In making her case for women's control of sexual selection, Gamble postulated that women could weed out characteristics detrimental to progress:

> The mother is the natural guardian and protector of offspring; therefore, so soon as women are free they will doubtless select for husbands only those men who, by their mental, moral, and physical endowments are fitted to become the fathers of their children. Only those women will become mothers who hope to secure to their offspring immunity from the giant evils with which society is afflicted. In this way, and this way only, may these evils be eradicated. (1894: 77–8)

Gamble here draws on her image of women's inherent altruism in making the case for local control of eugenics: individual women should be free to make choices, she implies, without the central coordination of the state. This would allow unfettered feminine altruism to weed out the 'evils' of modern human society as it always had in other species. In this way, Gamble and other Darwinian feminists embraced eugenics as a way to change human evolutionary essences over generations and improve the species. Gamble concludes that 'mankind will never advance to a higher plane of thinking and living until the restrictions upon the liberties of women have been entirely removed' (75). Providing women with scientific education was one immediate way Gamble felt society could accelerate the progress that had stalled under patriarchal capitalism.

Where Gamble's argument most closely resembles Blackwell's is in her description of scientific femininity. Like Blackwell, Gamble uses Darwin's list of essential female traits to elucidate why women are likely better natural scientists than men:

> Woman's rapid perceptions, and her intuitions, which in many instances amount almost to a second sight, indicate undeveloped genius, and partake largely of the nature of deductive reasoning; it is reasonable to suppose therefore that as soon as she is free, and has for a few generations enjoyed the advantages of more natural methods of education and training, and those better suited to the female constitution, she will be able to trace the various processes of induction by which she reaches her conclusions. (1894: 67)

Here Gamble points out that women's intelligence is not inferior to men's, but actually in the realm of genius. Like Blackwell, Gamble makes clear that women actually have the potential to become faster and better scientists than men, and all that is required is training. She not only makes a call for women to receive science education, but also to have it tailored to 'the female constitution' in a manner that was being developed at women's colleges such as Smith.

In *Descent of Man*, Darwin had singled out the brain and hand as two parts of human anatomy that had been developed in association with the development and use of technology, which was of central importance for human evolution (Sharp 2007: 42–7). He further attributed these developments to human males, but Gamble rejected this conclusion. Instead, Gamble claims that,

> Not only is man's sense of sight less perfectly developed than is woman's, but his sense of touch is less acute. The hand, directed as it is by the brain, is the most completely differentiated member of the human structure ... The female hand, however, is capable of delicate distinctions which the male has no means of determining. (1894: 50)

In other words, Gamble argues that women have both superior minds and hands for science. The two key physiological traits Darwin isolated as central to technomodernity and evolution – and that became the focus for the scientific extrapolations of men such as H. G. Wells (Sharp 2007: 70–85) – are here marked as a sign of female superiority.

While feminists certainly took the work of Spencer seriously, reform-minded feminists such as Blackwell and Gamble embraced Darwin's *Descent of Man* as a superior remedy to the religiously orthodox image of women from the Bible. Using Darwin's data and methodology – if not necessarily his interpretations and conclusions – Blackwell and Gamble helped forge a progressive Darwinian feminism that reached a wide number of people through their many publications and speeches. Blackwell was a close friend of women such as Susan B. Anthony and her many speeches to local, regional and national women's rights groups helped shape the generation that achieved women's suffrage in the United States. Despite Darwinian feminism's marginal position in the suffrage movement at the end of the nineteenth century, Gamble continued to speak and publish her ideas about evolution into the 1910s. *The Evolution of Woman* was widely reviewed, was 'favorably recommended in women's rights and freethought publications', and 'received positive reviews in the *Nation*, the *Critic*, and *Current Literature*' (Hamlin 2014: 141).

By the time Charlotte Perkins Gilman published her influential treatise *Women and Economics* in 1898, Darwinian feminist arguments had been circulating in progressive circles for over a quarter of a century. Gilman was a 'reform Darwinist' who 'considered evolution to be her religion', and used 'Darwinian evolutionary theory ... to probe the concept of the "natural"' (Hamlin 2014: 120). In *Women and Economics*, she also took Darwin's account of sexual selection to be an accurate diagnosis of an ill plaguing modern society:

> sex-distinction in the human race is so excessive as not only to affect injuriously its own purposes, but to check and pervert the progress of the race ... By the economic dependence of the human female upon the male, the balance of forces is altered. Natural selection no longer checks the action of sexual selection, but co-operates with it ... Man, as the feeder of woman, becomes the strongest modifying force in her economic condition. Under sexual selection the human creature is of course modified to its mate, as with all creatures. (1998: 19)

For Gilman, the bodies and aptitudes of men and women had been perverted by this unnatural dependency created by male control of sexual selection. Like Blackwell and Gamble, Gilman contends that only allowing women their economic freedom to compete on equal terms in education, politics and the work force would allow 'race progress' to continue along a better path.

However, what shape should this better path of progress for humanity actually take? Like Blackwell and Gamble, Gilman appealed to examples from non-human animals to diagnose the problems of human social relations and reimagine motherhood in natural terms. Gilman attacked the 'separate-spheres ideology' of her day and envisioned 'new domestic arrangements' based on animal models (Hamlin 2014: 122). Darwinian feminism thus provided the perfect springboard for the speculative futures and utopian visions of Gilman and many other women in the late 1800s and early 1900s. The critical edge of Darwin's work challenged the cultural dominance of the Eve myth and biblical arguments for gender relations. Conceiving of humanity as adaptable and capable of improvement, Gilman and her predecessors established plots, themes and concerns that came to pervade women's SF well into the 1930s.

3 Evolution's Amazons: Colonialism, Captivity and Liberation in Feminist Science Fiction

In a late chapter of *The Evolution of Woman* (1894), Darwinian feminist Eliza Burt Gamble examined and reinterpreted the role of Amazons in Johann Jakob Bachofen's *Das Mutterrecht* (1861) to show the natural fitness of women to be leaders in politics and society. In his multi-stage account of human history, Bachofen saw 'Amazonianism' arise in the first stage of matriarchal civilisation, die down with the full flowering of matriarchy and then re-emerge after the overthrow of matriarchy by patriarchy (Eller 2011: 43–5). While the idea of a society run by women certainly appealed to Gamble, the image of violent Amazons did not match her essentialist view that women are inherently sympathetic and nurturing. For Gamble, men are the violent and destructive sex and women are the cooperative builders of civilisation and carriers of progress. How, then, to explain violent Amazons? Gamble reasoned that,

> There seems to be considerable evidence going to prove that there have been times in the past history of the race in which women were brave in war and valiant in defending their rights ... Although the fact seems to be well substantiated that in certain portions of the earth, and at various periods in the history of the race, women have maintained their independence and protected their interests by force of arms, it seems quite as certain that actual warfare carried on by them has been confined to peoples among which male supremacy had but recently been gained, and among which a resort to arms represented the last act of desperation to which they were driven to maintain their dignity and honor. (1894: 183–4)

In order to further her argument about matriarchal society, Gamble had to separate the violent aspect of Amazons from their nurturing essences. In Gamble's interpretation, the Amazons are simply pushed to an unnatural extreme in order to protect their 'dignity and honor' and maintain control of their reproductive capacities. Gamble goes on to conclude that 'these cases have been exceptional' and that there is 'no cause for thinking ... armed resistance to masculine authority has been uniform or protracted among' women (1894: 184). By representing female violence as an extreme exception, Gamble was consistent with a long history in America of representing women's violence as emerging only when threatened by colonisation and rape. This American reinterpretation of the Amazon figure deviated from the classic Greek model.

As Batya Weinbaum argues, the figure of the Amazon emerged from an ancient, pre-Homeric Greek society that repeatedly played out 'the psychodrama' of a matriarchal society confronting 'the culture of the newly emerging city-state of Greek patriarchy' (1999: 79–80). The primary purpose of such stories seemed to be 'to teach Athenians and other "civilized" peoples that the rule of women was freakish, dangerous, and certainly not to be risked in any form' (Eller 2011: 17). The most famous Greek account of Amazons came from the historian Herodotus in the fifth century BC. According to Herodotus, a group of Amazon warriors were captured by the Greeks. While sailing back to Greece, the captive Amazons on 'three of their vessels' eventually 'rose up against the crews, and massacred them to a man' (Herodotus 1972: 28). After landing ashore in a strange country of Scythians, they began raiding to stay alive. The Scythians sent 'a detachment of their youngest men' to live near the Amazons and woo them (29). After they successfully began 'living with the Amazons as their wives' the young Scythian men asked them to return home (29). However, the Amazons refused, saying that, 'We could not live with your women – our customs are quite different from theirs. To draw the bow, to hurl the javelin, to bestride the horse, these are our arts – of womanly employments we know nothing' (29–30). They despised the domestic lives led by the Scythian women. Instead of moving to Scythian homes, the Amazons convinced the men to join them in moving to a new area. Herodotus concludes by saying that 'women of the Sauromatae have continued from that day to the present to observe their ancient customs, frequently hunting on horseback with their husbands, sometimes even unaccompanied; in war taking the field; and wearing the very same dress as the men' (30). Through his tale, Herodotus demonstrates how strong women are incompatible with Greek society, and can only be assimilated into the culture of barbarians such as the Scythians, and even then only with some compromises. This association of Amazons with supposedly uncivilised cultures became a central aspect of the stories of 'medieval Europeans', and the stories of 'conquistadores frequently claimed that they encountered Amazons in the lands they explored' (Eller 2011: 20). In fact, the first wave of colonists in the Americas seemed obsessed with finding Amazon tribes in the new world. In the 1540s, Francisco Pizarro sailed through a river in South America and claimed that he and his men 'were attacked by powerfully built women warriors, from all-female tribes in the area', thus leading to the river being named the Amazon (Moskowitz 1972: 3).

Amazons in America

Amazons became domesticated in early American colonial literature as defenders of nascent Euro-American civilisation instead of the earlier association of Amazons with barbarism or savagery. While women were excluded from the institutions of science and rarely went on voyages of discovery, they played an important role in Europe's colonies in the Americas. In the evolving system of genres that circulated between Europe and the colonies, the captivity genre became one site where strong women began to have a significant presence. Scholarship on American captivity narratives involving women captives has shown the complexity of such stories in regard to the reification of – and resistance to – contemporary beliefs about gender. While many such narratives emphasised the passivity of the woman captive, some also included a critique of the 'patriarchal constraints that keep [women] captive once they've returned home' (Tinnemeyer 2006: xvi). These American captivity narratives relied on the dramatic tension brought about when a racially and religiously 'other' male threatened the white female captive. Captivity narratives thus provided one way for the American colonists to 'struggle through questions of cultural and gender identity during periods of extreme change and uncertainty' (Namias 1993: 10–11). For the American Puritans, captivity was characterised as a 'God-given affliction designed to chastise the victim into moral and spiritual reformation' (Colley 2003: 202). Consequently, Puritan writers used captivity narrative as a way for their heroes to complete a 'heroic quest' that included 'religious conversion and salvation' (Slotkin 1996:

21). These early European colonists were like other Europeans of the late seventeenth century in that religion was the main difference they saw between themselves and the 'savages' they encountered. The early American captivity narratives produced by colonists often served to reinforce distinct gender norms deemed appropriate by the authors. In what June Namias calls 'American Amazon' narratives, however, the realities of colonial life enabled women to step outside accepted gender roles because of the absence of men. As Namias observes, 'Husbands went away on long trips; husbands died. In such circumstances, society sanctioned women to serve in their stead for the sake of family survival' (1993: 29).

Hannah Dunstan's experience provided the material for the prototype American Amazon story that was rewritten several times over the following two centuries. The earliest accounts of Dunstan's experience were published by the Puritan minister Cotton Mather. Like the Puritan captivity narratives that had preceded it, the complex of genres in Mather's published 1702 account drew heavily from the format and rhetoric of the Puritan sermon. Cotton Mather's father Increase had helped publish the famous 1682 account of Mary Rowlandson's captivity that had been the seventeenth-century equivalent of a bestseller in England and its American colonies (Namias 1993: 9). A preface to Rowlandson's story claimed that *This Narrative was Penned by this Gentlewoman her self* (Rowlandson: 1998: 9); however, Dunstan's story came exclusively from the pen of Cotton Mather. With his control of the story, Mather crafted an account of a strong woman whose heroism is seen as acceptable only within exceptional circumstances, whose strength comes from religion and frustrated motherhood, and who safely returns to the domestic sphere.

Dunstan was taken captive by a group of Abenaki in 1697 during King William's War (Namias 1993: 29–30). As Mather's account begins, Dunstan is recovering from childbirth when a group of '*Salvages*' (savages) attack Haverhill, Massachusetts (Mather 1998: 58). Dunstan's husband returned 'from his Employments abroad' but could only rescue seven of their eight children in time (58). Dunstan, her nurse and her newborn are captured. As the 'furious Tawnies' march Dunstan away from her home with several other captives, her child is murdered when their captors 'dash'd out the Brains of the *Infant* against a Tree' (58–9). Mather takes time to catalogue the brutality of the attack, lamenting the fate of those captives who begin 'to Tire in their sad

Journey' away from Haverhill: 'the *Salvages* would presently Bury their Hatchets into their Brains, and leave their Carcasses on the Ground for Birds and Beasts to Feed upon' (59). In many ways, this is a standard colonial representation of Native Americans as bloodthirsty killers. In the nineteenth century, such representations would link the cruelty of Native Americans to their race: they were seen as simply behaving according to the dictates of their biological essence. However, in this period before scientific notions of race were fully developed, what supposedly leads the 'savages' to behave so cruelly is their lack of Christianity.

This can be seen most clearly in Mather's account of Dunstan's long journey with her Abenaki captors. Mather describes the many physical hardships Dunstan endures only briefly, but spends a substantial amount of time describing the religious persecution Dunstan and her nurse go through at the hands of 'those whose *Tender Mercies are Cruelties*' (59). Mather turns Dunstan's faithfulness in the face of persecution into an object lesson for the '*English Family*, that has the Character of *Prayerless* upon it' (59). Mather says that the Abenaki captors were '*Idolaters*' who 'were like the rest of their whiter Brethren *Persecutors*, and would not endure that these poor Women should retire to their *English Prayers*, if they could hinder them' (59). Mather praises Dunstan and her nurse for finding ways to pray in secret. True to the sermon genre, he also shames those 'prayerless' families who do not appreciate their freedom to pray without molestation.

Dunstan's status as a pious victim is consistent with idealised images of women of the period: like Mary Rowlandson, she is patient, faithful and longsuffering in her devotion. However, many details of the story defy the image of a passive religious woman. Dunstan and her nurse make it through 'Travel of an Hundred and Fifty Miles ... within a few Days Ensuing, without any sensible Damage in their Health' (59). Dunstan and her nurse are clearly seen as hearty frontierswomen who survive many hardships that break other captives. Eventually, the spectre of physical violation spurs Dunstan and her nurse to action. As Namias points out, the sexual threat posed by Indian men in captivity narratives often ended in gruesome scenes of their murder and dismemberment by white women (1993: 94). The sexual nature of the threat is often oblique, but in Mather's account the threat of humiliation and violation is clear: when they reach their destination, Dunstan and her nurse would 'be Stript, and Scourg'd, and Run the *Gantlet* through the whole Army of *Indians*' (1998: 60). Immediately after this threat is presented in Mather's text, he recounts Dunstan's turn to violence:

> about an Hundred and Fifty Miles from the *Indian* Town, a little before break of Day, when the whole Crew was in a *Dead Sleep* ... one of these Women took up a Resolution to intimate [sic] the Action of *Jael* upon *Sisera*; and being where she had not her own *Life* by any *Law* unto her, she thought she was not forbidden by any *Law* to take away the *Life* of the *Murderers*, by whom her *Child* had been Butchered. (60)

In Mather's hands, Dunstan's story becomes one of tortured motherhood and bloody retribution for her lost child. He emphasises the common colonial argument that the laws of civilised Christian society do not apply when dealing with '*Salvages*' (savages) and '*Idolaters*'. In the process, Dunstan is represented as completely justified in getting two other captives – her nurse and a young boy – to help her kill and scalp ten of the twelve members of the Indian party with which she was travelling (most of whom were women and children) (Namias 1993: 30).

Like the explorers and scientists of the period, Dunstan became hailed as a champion of colonial progress. Dunstan's story highlights the kinds of acceptable violations of gender boundaries that began to circulate through colonial genres in the seventeenth century. As a woman, Dunstan is celebrated in the text for engaging in very unwomanly violence when confronted with violation by savage men. Because they were enemies during a time of warfare and because they lived outside the Christian laws of Puritan society, Mather asserts that the Abenaki Indians could be killed with no danger to one's immortal soul. Indeed, Mather ends Dunstan's story with a brief account of the congratulations and monetary rewards she received as a result of her experiences and actions. Interestingly, the biblical model Mather evokes for Dunstan's actions is Jael, a Kenite woman in the Book of Judges who murdered an enemy of Israel named Sisera in his sleep after a battle. Jael would later provide the name for classic feminist SF characters in *The Female Man* by Joanna Russ (1975) and *The Shattered Chain* by Marion Zimmer Bradley (1976). Like the biblical Jael, Dunstan serves as a woman whose violence is both justified and necessitated by the uncertainties and realities of warfare.

At the same time, the subversive nature of Dunstan's actions is contained in many ways. Mather represents her violence as motivated (at least in part) by her status as a mother: she saw her newborn murdered, and revisits that violence upon her child's murderers. The dictates and emotions of motherhood were to become a common feature of stories about violent women in the centuries following Dunstan's narrative. In this sense, Dunstan's story reinforces the importance of motherhood to female identity. The threat posed to Dunstan by Indian men also serves as a gall to the Puritan men encountering her story. If Christian men had adequately carried out their role as protectors of their women, Dunstan would not have had to resort to violence to protect her modesty. The absence of Christian men allows Dunstan's character the space she needs for subversive violence, but ultimately this is contained by her return to civilisation. In Dunstan's captivity narrative – and many that came after – the story ends with the captive women returning to their 'natural' place in the domestic sphere (Tinnemeyer 2006: 88–9). Cotton Mather's control of the story ensures that its subversive elements are safely subordinated to a Puritan worldview, and Dunstan's captivity narrative ultimately served the imperatives of empire and male dominance.

Popular Amazon captivity narratives like Dunstan's re-emerged during war crises throughout the eighteenth and nineteenth centuries. As the genre developed in America, it often combined with a long-standing genre from Europe that depicted women dressing up as soldiers and fighting in combat like men. As with American Amazon captivity narratives, most stories of the woman soldier in drag also played on the sensational image of a woman defying traditional gender roles. They also took the actual experiences of tough and transgressive American women and turned them to the purposes of colonialism and patriarchy. In the fictional (or quasi-fictional) stories that became popular in the nineteenth century, the women soldiers would join up with the army to pursue or avenge some male lover or relative. Much of the tension in the stories came from the threat of their sex being discovered, with many falling clearly into a particular type of captivity narrative. One example is the anonymously authored *The Female Warrior: An Interesting Narrative of the Sufferings and Singular and Surprising Adventures of Miss Leonora Siddons*, a first-person story that was published in 1843 during the ongoing struggle to annex Texas from Mexico. In the story, Siddons is a Texan whose father dies fighting the Mexicans: 'Without a friend or protector', the orphaned Siddons heeds her father's words to serve her country and runs off to join the fight against Santa Anna disguised as a male (Anon. 2007: 6). In the account of the main battle at San Antonio, Siddons repeats the tone and rhetoric of pro-American war propaganda of the period. For example, the Mexicans outnumber the Texans 'ten to one' just as at the Alamo, which was still fresh in the minds of the reading audience (8). The noble Texans are eventually overwhelmed by sheer numbers and Siddons is captured only after fighting for hours. When a large Mexican puts his hand on her and claims her as a prisoner, she attempts to fight to the death out of a fear of sexual violation:

> 'Never!' cried I, as I thought of my sex. 'Liberty or death!' and as I spoke I made a pass at him with my cutlass, which I still held in my hand.
> This he was prepared for: catching it on his with a quick dexterous turn, he wrenched it completely from my grasp.
> 'Lost, lost, all is lost!' exclaimed I, in agony; then perceiving a brace of pistols in his belt, I sprang with the desperation of a madman, and, ere he was aware of my intentions, drew one, and shot him through the heart.
> At that instant I saw the flash of another, and all was dark – I fell. (9)

At this point, the narrative clearly draws on the conventions of captivity narrative: the woman is captured by enemy others who pose a sexual threat. In one sense, Siddons resembles the type of stock captivity character that Namias calls the 'Frail Flower', or the hysterical, fainting, woe-is-me damsel who needs to be rescued (Namias 1993: 36–7). She wails 'all is lost' and her behaviour is described as 'mad' before she faints. The twist, of course, is that Siddons's hysteria is mixed with violent bravery, and her 'fainting' is precipitated by a bullet grazing her skull. In this sense, Siddons is both a brave Amazon and a frail flower, a curious hybrid that can be called the 'Amazonian flower'. She is menaced by her racially other captors, but she is protected against rape by her male drag. She is a flower in soldier's clothing, and it is her attempt to maintain her secret feminine weakness that drives the narrative. Like the American Amazon, she kills racial others who threaten her, but like the frail flower she collapses and ultimately needs the help of men to escape. The Amazonian flower became a stock figure in adventure fiction and SF written by men in the twentieth century.

Amazons and Darwinian feminism: Mary Bradley Lane's *Mizora*

Feminists of the late nineteenth century took the image of the Amazon in another direction, one that rejected the imperatives of patriarchy while maintaining the colonial dominance of whites over other races. In a series of utopian stories, feminists depicted Amazonian societies where men are either absent or destroyed in order to liberate women from their oppression. While these stories jettisoned much of the ideological baggage of the American Amazon, they tended to retain a colonial perspective in imagining their utopian civilisations as based upon superior European bloodlines and by locating them beyond the most geographically remote 'savage' societies. They also expanded the utopian impulses of Darwinian feminism into fully fleshed out societies that imagined how women could evolve beyond the limitations set by patriarchy. The earliest of these was Mary Bradley Lane's *Mizora: A World of Women*, which was serialised in four parts for the *Cincinnati Commercial* newspaper in 1880 and 1881 and published as a book in 1889. Lane was a former public schoolteacher from Ohio who hid her publication and radical politics from her husband (Fisher 2014: 181–2). Her story was consistent with feminist debates of the day, particularly in regard to sexual selection, education and scientific femininity. Though there is no direct evidence, it is plausible that she was aware of the work of Antoinette Brown Blackwell, a noted Darwinian feminist who went to college in Ohio and whose speaking tours and varied publications gave her arguments about marriage, evolution and women in science a wide audience (Fisher 2014: 185). *Mizora* includes a number of ideas that are consistent with Blackwell's, with Lane expanding and embroidering them through her utopian society of Amazons.

In some respects, the plot of Lane's *Mizora* resembled English novelist Edward Bulwer-Lytton's popular novel *The Coming Race* (1871), which featured a traveller visiting an ancient society that had evolved underground and that sees women as equals or superiors to men in many ways. However, in Lane's story the visitor is a woman instead of a man, and men have mysteriously vanished, a mystery that hangs over the first half of the text. Lane also seems to have drawn elements from Russian socialist utopian writing exemplified in Nikolai Chernyshevsky's *What is to Be Done? Tales about New People* (1863), particularly in her embrace of 'mutual aid that dominated the adaptation of Darwin's ideas for Russian society' (Fisher 2014: 195). Lane's story broke new ground for Darwinian feminists, incorporating these many influences into a form that anticipated and inspired feminist SF for the next century. The narrator of *Mizora* is named Vera Zarovitch, a wealthy Russian woman who is forced to flee her country because of her 'sympathy' for a 'Polish orphan' and her 'oppressed people' (Lane 1999: 9). This sympathy leads her to insult her government when she witnesses Russian soldiers committing an atrocity. Vera shares a first name with a major character in Chernyshevsky's novel, and her depiction of the tyranny of Czarist Russia shows Lane's familiarity with socialist arguments regarding Russian politics (Fisher 2014: 196–8).

Despite these Russian influences, Lane's use of a frontier setting shows the peculiarly American flavour of her socialist story. After escaping prison through bribery, Vera heads north 'to the frontier' aboard a 'whaling vessel bound for the Northern Seas' (1999: 11). The unscrupulous crew abandons her with a group of 'Esquimaux', but Vera quickly adapts by wearing a 'suit of reindeer fur' and eating 'the raw flesh and fat' offered by her hosts (11). Vera notes that she 'cheerfully assumed a share of their hardships, for with these poor children of the North life is a continual struggle with cold and starvation' (11). Vera travels with them until they come across open sea facing north. When she expresses her desire to sail across that sea, her 'friend' warns her, 'Across *that* no white man's foot has ever stepped' (12). In Vera's trip north, Lane evokes the frontier toughness of Hannah Dunstan and anticipates the Darwinian struggle to survive on the frontier that would become a hallmark of the historical writing of both Theodore Roosevelt and Frederick Jackson Turner (Sharp 2010). Where Roosevelt and Turner associated such survival with manliness, Lane shows a strong heroine able to rough it in the worst 'savage' conditions nature has to offer. Where Dunstan's story emphasised fighting the natives, Vera instead joins them and shows sympathy for their lives. She is not taken captive, but instead is welcomed like a friend. Lane adopts a peaceful approach to discovery, and forgoes the violent conflict of colonial American Amazon narratives and masculinist Darwinian accounts of colonisation.

When Vera launches further north in a boat, Lane draws on popular contemporary accounts of voyages attempting to find the Northwest Passage that taught readers 'lessons about national character and traits of manliness' (Robinson 2015: 91). Where men had failed to find the route through the Arctic sea, Vera succeeds and is pulled down into an underground world. In this world, she comes across a perfect society that allows her to teach the reader lessons about women's oppression and the possibilities offered by their emancipation. Upon landing in Mizora – the name of the underground country she discovers – Vera observes that there are only women. Late in the story, 'The Preceptress' of Mizora's 'National College' explains that there are no more men, and that the women of Mizora have learned to control 'the germ of all Life' (Lane 1999: 103). Lane's portrait of ancient Mizoran history mirrors the history of the United States, where repeated wars and a final civil war 'severed forever the fetters of the slave and was the primary cause of the extinction of the male race' (96). When the government began to collapse, the women 'organized for mutual protection from the lawlessness that prevailed. The organizations grew, united, and developed into a military power. They used their power wisely, discreetly and effectively. With consummate skill and energy they gathered the reins of Government into their own hands' (100). As in early American Amazon narratives (and Gamble's later account of Amazonian societies), the Mizoran women only became military Amazons to protect themselves from male 'lawlessness', violence and domination. In due course this led Mizoran women to the holy grail of Darwinian feminists: control of sexual selection. With women scientists able to control the germ of life, they are able to reproduce without men through a vague kind of parthenogenesis. There is no romance in Mizora, and with women firmly in control of their bodies they become chaste and completely devoid of the animal sexual desire that Blackwell and other Darwinian feminists had attributed solely to men. Though rejecting male control of sexual selection, Lane's Amazons still remain locked within a heterosexual economy of desire: the only possibilities presented are sex with men or sexless reproduction, and sex itself is only conceived of in terms of reproduction. Same-sex desire remains excluded as a possibility.

Like many of her feminist contemporaries, Lane marks sexless motherhood as centrally important to female identity in her utopian society (Roberts 1993: 73). Vera reveals that the 'only intense feeling that I could discover among these people was the love between parent and child' (Lane 1999: 32). An important public manifestation of this love is education, and it is a top priority for Mizorans 'to secure the finest talent for educational purposes' and pay them better than anyone else (23). For a former schoolteacher such as Lane, the negative effects of underpaying educators were familiar terrain. Before the women's revolt in Mizora,

> They had been hampered in educational progress. Colleges and all avenues to higher intellectual development had been rigorously closed against them. The professional pursuits of life were denied them. But a few, with sublime courage and energy, had forced their way into them amid the revilings of some of their own sex and opposition of the men. It was these brave spirits who had earned their liberal cultivation with so much difficulty, that had organized and directed the new power. (100)

Educated feminists are therefore the revolutionaries who led the foundation of Mizora's matriarchy in the face of harsh resistance from men and some women. Lane then extends the ideal of motherhood into a fantastic expansion of the domestic sphere where the entire society is organised like a well-ordered home and 'The State was the beneficent mother who furnished everything' for her citizens (23). With women in control of both sexual selection and the reins of government, Lane posits that a nurturing socialist state is truly possible.

For Lane, the key to feminist emancipation and the continued health of women is unfettered access to education. Lane comments on this through the Mizoran belief that failure to make education free for all would lead them to 'relax into ignorance, and end in demoralization' (24). Lane's ideas about education serve as an endorsement of the Morrill Act of 1862, which created land grant colleges in the United States and made education more accessible for men and women (Rossiter 1982: 67). Like these emerging land grant colleges, Lane seems to have a special affection for the natural sciences. Following in the footsteps of Blackwell, Lane puts forward a vision of scientific femininity that is less concerned with warfare and more concerned with transforming everyday life in ways that liberate women from drudgery. The Preceptress explains to Vera that, 'Science had hitherto been, save by a *very* few, an untrodden field to women; but the encouragement and rare facilities offered soon revealed latent talent that developed rapidly' (1999: 103). With 'every state' creating a 'free college provided for out of the State funds', women have the chance to work in 'every department of Science, Art, or Mechanics' which are 'furnished with all the facilities for thorough instruction' (23). The products of this feminine progress far outstrip the achievements of the United States of the nineteenth century: Mizorans have created artificial rain and a horseless carriage. Every home also has communication devices where a person from far away 'appeared upon the polished metal surface like the figure in a mirror and spoke' to the person who contacted her (78). Such inventions demonstrate how Mizoran science values connection and community over isolation and conquest (Roberts 1993: 72–3). In this way, Mizoran science fulfils the claims of feminists such as Blackwell, who argued that women were better suited for science and civilisation than men. It also anticipates the rise of home economics between the 1890s and 1910, a field dominated by women based in analytic chemistry and biochemistry devoted to providing practical advice for women and helping with 'public health problems' such as malnutrition and child mortality among poor and immigrant populations (Rossiter 1982: 66).

Despite the feminine focus of Mizoran science, Lane still evokes the language of colonial conquest to describe their feats. Vera explains that 'they had laid their hands upon the beautiful and compelled nature to reveal to them the secret of its formation' (1999: 48). Though this rhetoric is used to describe the process of discovery, the rhetoric shifts when it comes to applications. Repeatedly the women of Mizora emphasise how they 'follow the gentle guidance of our mother, Nature' (105). Mizoran science created a 'feminized science of domestic culture' that liberates all women from the worst aspects of domestic work (Fisher 2014: 193). Vera is amused by 'a little machine, with brushes and sponges attached, going over the floor at a swift rate, scouring and sponging dry as it went' (Lane 1999: 44). Vera concludes that Mizora is 'a land of brain workers. In every vocation of life machinery was called upon to perform the arduous physical labor. The whole domestic department was a marvel of ingenious mechanical contrivances. Dishwashing, scouring and cleaning of every description were done by machinery' (45). The elimination of domestic labour is replaced with an artisan-like devotion to domestic science. Chemistry was a particular love of Mizora, where they engage in 'the chemical production of bread and a preparation resembling meat' (26). Instead of being considered servants, cooks are chemists who 'can put together with such nicety and chemical skill the elements that form an omelette or a custard' (36). In this way Mizoran women elevate what was domestic drudgery – cooking meals as a servant – into a highly valued art. Their conquest is not of other peoples or of nature, but of the things that make daily life miserable for women.

Perhaps the most troubling aspect of Lane's utopia is the extreme form of eugenics practised by the Mizorans. When she arrives, Vera describes the Mizorans as 'the highest type of blonde beauty' that contrasts with her own dark hair and eyes (15). Later in the novel, the Preceptress takes Vera to study ancient portraits of the men who used to rule. Vera notices portraits of men who have dark hair and eyes, and the Preceptress informs her that Mizorans 'believe that the highest excellence of moral and mental character is alone attainable by a fair race. The elements of evil belong to the dark race' (92). Vera, however, does not accept this, noting to herself that 'their admirable system of government, social and political, and their encouragement and provision for universal culture of so high an order, had more to do with the formation of superlative character than the elimination of the dark complexion' (93). Here Lane shows ambivalence toward the white purity of the Mizorans, and shows sympathy toward 'the manipulated and rejected objects of science' she resembles (Fisher 2014: 193). Pushing back against the biological determinism of the Mizorans – and by implication, some of her fellow feminists – Lane promotes a Lamarckian understanding of inheritance where characteristics acquired or developed by individuals can be passed on biologically to their offspring. In other words, Lane believes better institutions can drive biological progress and implies that it has nothing to do with race. While her image of superior humans is a blonde one, Lane hedges on the desirability of such a fair white future in a way that anticipates the work of women such as Lilith Lorraine in the late 1920s and 1930s.

While ambivalent about the importance of race, Lane makes clear her support of eugenics for achieving a feminist future. In her discussion with Vera, the Preceptress gives a lecture on eugenics that echoes Blackwell's account of nature: 'We have advanced enough in Science to control its development. Know that the MOTHER is the only important part of all life. In the lower organisms no other sex is apparent'

(Lane 1999: 103). The Preceptress continues, 'By the careful study of, and adherence to, Nature's laws' Mizorans were able to eliminate 'the courser nature of men' over the centuries (104). With complete control of reproduction, and without men and their inferior traits, these super mothers 'control Nature's processes of development' at will (104). This leads to the elimination of all violence and criminal behaviour through negative eugenics. Both Blackwell and Gamble asserted progress was impeded by the violent, regressive tendencies that were engrained in the biological makeup of men. Lane shows the progress that could result from giving mothers complete control of sexual selection: it could finally lead to the elimination of the worst aspects of humanity by making men biologically obsolete. It would allow for the order of domesticity to spread across entire societies, and prevent women from having to cope with the degrading animal desires of men.

Amazons, utopias and adventure

Feminist utopias were not only confined to the United States in the late nineteenth century. British feminist Elizabeth Burgoyne Corbett's *New Amazonia* (1889) was deeply influenced by Francis Galton's new ideas about eugenics, and shared a number of characteristics with its counterparts from the other side of the Atlantic. *New Amazonia* imagined a utopian world led by women from England who colonise a ravaged Ireland. The novel has a largely satirical tone, with most of its humour aimed at British manhood through a character Augustus FitzMusicus who the narrator describes as 'a perfect specimen of the British masher' (2014: 35). The narrator, a nameless British woman who is suspiciously similar to the author, becomes transported into the future with the hapless FitzMusicus after reading an article written by 'ladies' against women's suffrage. In the prologue, the narrator rails against such 'ladies' as 'frivolous' and materialistic, concluding that they are prone to 'despise and depreciate every woman who recognises a nobler aim in life than that of populating the world with offspring as imbecile as herself' (26). The narrator allies herself with *'women'* who 'want something more substantial' and who 'prize independence above all things' (28). The prologue establishes Corbett's critique of unthinking motherhood, the solution to which is found in feminist eugenics.

Corbett expands upon the possibilities for progress that come from women's enfranchisement in a manner similar to Blackwell and Lane. Once transported to the future the narrator and FitzMusicus are taken in by giant Amazons who are the descendants of English women who have left behind the old 'warlike race' of England dominated by men (2014: 39). While studying the history of New Amazonia, the narrator learns that 'Universal Suffrage' in England allowed women to show 'themselves so much more just, and so much more capable of governing than men' (64). War had depleted the male population of Britain, and it is decided 'to colonise Ireland with the women who outnumbered the men so enormously' (65). Corbett draws on the Darwinian formulation of women as more selfless than men, with their concerns turning to building a superior civilisation rather than expanding their territory beyond Ireland. In the society they build, 'the State was the Mother of her people', and provided for or managed almost every aspect of life (86). The rhetoric used to explain this expansion of the domestic sphere to the level of the state is nearly identical to Lane's *Mizora*, and would be repeated by a number of other feminists over the years. However, Corbett does not represent an egalitarian state: women who wished to immigrate to the new country had to 'invest a certain sum of money', and any 'woman who bore the slightest trace of disease or malformation about her was rigorously rejected' (68). In other words, New Amazonia was based on a strict, government-mandated policy of negative eugenics that also privileged middle-class and upper-class women. This differs from the decentralised vision of positive eugenics embraced by some socialist feminists, who believed that emancipated women would naturally make the best choices about reproduction without the need for government intrusion.

The scientific women of *New Amazonia* show a similar focus on medical and domestic technologies to those from *Mizora*. The meals of New Amazonia are vegetarian and 'scientifically perfect', with the food 'of such a nature that it at once supplied all the wants of the body' (2014: 88). In her descriptions of their scientific achievements, the narrator uses the familiar rhetoric of conquest: 'Electricity was made so thoroughly subservient to human will that it supplied light, heat, and powers of volition, besides being made to perform nearly every conceivable domestic use' (90). The domination of nature has eliminated smoke and other pollution, and a system of 'hydraulic roads' allows quick and easy movement throughout the country (92). Like

Lane's *Mizora*, scientific femininity dominates nature for the health of people in an expanded domestic sphere, not for the domination of people outside the bounds of the country. That said, it takes the Irish a while to accept their new colonial masters and 'take kindly to the mode of living universally enforced throughout the country' (70). In their initial founding act of colonisation, these Amazons are also decidedly non-violent, though Corbett emphasises their capacity for violence in a way that most other feminist utopias do not. The narrator discovers that 'standing armies were seldom maintained', but is informed that New Amazonians 'are all trained to fight, and there is not a woman or man in the country who does not thoroughly understand military discipline. Our training begins in infancy, and includes riding, shooting, swimming, diving, ballooning and every possible military exercise' (119). Corbett retains the fierceness of Amazon warriors as a necessary protection for the utopian world they have built. This military training is rarely represented in American feminist utopias of the period. However, Corbett does retain the defensive nature of this training, preserving the image of female violence as an exception that only emerges when threatened by colonisation or male domination. She also embraces the need for men in reproduction, with science only ensuring eugenically desirable mates instead of replacing them altogether.

Not all utopian worlds created by women in this period were focused on Darwinian feminist ideas. Pauline Hopkins's *Of One Blood; Or, the Hidden Self* (1903) includes a utopian civilisation that draws a sharp contrast to those of Darwinian feminists, using the Bible as a foundation for extrapolation and embracing male power in an ideal world. Hopkins was an African American author from Boston who edited the *Colored American Magazine*, where she published *Of One Blood* and two other novels in serial format (Carby 1988: xxix–xxx). *Of One Blood* follows a mixed-race protagonist named Reuel Briggs who is passing as white while attending university and who is 'a genius in his scientific studies' (Hopkins 1988: 444). Through a series of traumatic experiences, he eventually ends up joining an expedition to Ethiopia to study the ruins of an ancient civilisation. The novel draws elements from popular story types such as the Edisonade, which follow the exploits of Thomas Edison-like heroic scientists, and 'the lost-race adventure' where explorers discover a previously unknown ancient culture or branch of humanity (Williams 2012: 323). However, Hopkins spins these story types in a manner consistent with what is now called

Afrofuturism: the genius scientist is African American and the lost race turns out to be an advanced Ethiopian society poised to dominate the white world with Reuel as its leader. Where Darwinian feminist utopias up to this point tended to be based on Aryan motherhood and European civilisation, Hopkins 'portrays Africa as the cradle of an ancient black civilization that pre-dated white civilizations' and that is still vibrant (Ficke 2016: 66). Reuel's mother is a former slave, and in learning to embrace his African heritage he also recovers his connection to his mother (Bergman 2008: 294). However, the utopia he discovers is not matriarchal, as its queen only rules while she awaits his long-prophesied arrival. Throughout Hopkins's novel, evolution is rejected and the Bible is 'reestablished as authority through technologically-enhanced exploration' (Williams 2012: 324).

Less than a decade later, Edgar Rice Burroughs sparked a virulently masculinist embrace of Darwinian beliefs about gender in his adventure stories *Under the Moons of Mars* (1912) and *Tarzan of the Apes* (1912), both of which were published serially in *The All-Story* magazine. Both were fantastically successful, spawning several sequels, numerous films and countless imitators. His heroes, John Carter and Tarzan, were an idealised blending of civilised manners and primitive violence, and their stories dramatised repeatedly their mastery over women, non-human animals and non-white or alien races (Sharp 2007: 92–6). When confronted by these ideal men, women tended to swoon and be moved to ardent passion. For example, when Tarzan rescues Jane Porter from a giant ape-like creature in *Tarzan of the Apes*, she becomes 'a primeval woman who sprang forward with outstretched arms toward the primeval man who had fought for her and won her' (Burroughs 1997: 137). Burroughs designed female characters as beautiful prizes to be won by real men, and their only real purpose is to get into trouble so that they can be rescued. Ironically, Burroughs and his Darwinist adventure yarns became a primary influence on the most prolific women writers of SF of the late 1920s and 1930s.

Feminist angels: Inez Haynes Gillmore's *Angel Island*

In the late nineteenth century, the Victorian image of the 'angel in the house' became the subject of particular scorn for feminists. In her landmark 1898 treatise *Women and Economics*, Charlotte Perkins

Gilman argued that this economic dependence exaggerated differences between the sexes in a way that was injurious to humanity. The eugenic vision of Gilman's *Women and Economics* became the basis for her famous 1915 utopian novel *Herland* (Donawerth 1997: 20; Hausman 1998: 491-3). However, Gilman's *Herland* was likely influenced by a similar novel entitled *Angel Island* (1914) that was published a year earlier by her friend and sister feminist Inez Haynes Gillmore. An alumna of Radcliffe College, Gillmore became friends with Gilman at the Heterodoxy Club, a Greenwich Village establishment that opened in 1912 and catered to leading feminists of the day. *Angel Island* was published serially in *American* magazine, and was released later in 1914 as a book (Rich 2004: 155-7). Gillmore was a founding member and fiction editor of the radical leftist magazine *The Masses* that began publication in 1911, and authored several books including many under the name Inez Haynes Irwin after her remarriage in 1916 (Jones 1993: 1-3). Gillmore had been a labour journalist before working on *The Masses*, and became a 'well-known suffrage activist and historian' as well as 'a member of the National Advisory Council of the National Woman's Party on the eve of the suffrage victory' (137). *Angel Island* was republished in the February 1949 issue of *Famous Fantastic Mysteries* by editor Mary Gnaedinger, indicating that succeeding generations of feminist SF authors were familiar with Gillmore's work despite the fact that the story is rarely mentioned in histories of the genre (Davin 2006: 231-2).

Angel Island resembles the lost race story of Lane's *Mizora*, but differs significantly from earlier feminist utopias in that much of it is narrated from a male perspective and takes on a dystopian feel halfway through the novel. It opens with the aftermath of a shipwreck where five men have washed ashore on a remote island in the Pacific. Each of the men represents a different type of manhood that has a different attitude toward women, and through her descriptions of them Gillmore also assigns each of them to a different racial type. What they share is a primal, evolutionary masculinity that is buried beneath the civilisation from which they are now isolated. This is first noticed by Frank Merrill, the sociologist professor who says to his shipwrecked companions, 'you can thank whatever instinct that's kept you all in training. This shipwreck is the most perfect case I've ever seen of the survival of the fittest' (Gillmore 1988: 28). The fact the men survived the shipwreck is represented as a return to natural selection, where the weakening influence of civilisation is stripped away and only the strongest men survive. It also evokes the model of primitive masculinity championed by Gillmore's contemporaries Theodore Roosevelt and Edgar Rice Burroughs.

The angels of the title are five winged women who are described by one of the men as, 'members of a lost species – the missing link between bird and man' (205). They are 'discovered' by the men after their shipwreck, though it is clear from the novel that the women discover the men as much as the other way around. Through the eyes of these men, Gillmore playfully mocks masculinist fantasies about the lost race of female 'angels' and the society the men imagine they will build with them. Billy Fairfax, a big wealthy 'shockheaded blond', dreams about 'a mixture of "Arabian Nights," "Gulliver's Travels," "Peter Wilkins" ... Jules Verne, H. G. Wells, and every dime novel I've ever read' (33, 67). As they discuss the possibilities together, the five men imagine 'Angel Island would be an Atlantis, an Eden, an Arden, an Arcadia, a Utopia' (68–9). With five men and five angels, they begin to pair up in an extended flirtation that goes on for several chapters. Tortured by the flirtatious women hovering out of their reach, the men build homes and a small community. However, the stress of the island and the frustration of the men eventually turns to a brutal discussion about the possibility of 'marriage by capture', something that is associated with 'the principles' of 'Hottentots and Apaches and cave-men' (130–1). One by one each of the men is overcome by 'a feeling or an instinct' that is 'too strong' to fight (150). Here Gillmore repeats the common association of marriage by capture with supposedly primitive races, a problematic attribution that was central to the Darwinian feminism of the day. By doing so, she also evokes the Darwinist formulation of the brutal, violent male who uses force to take control of sexual selection.

Through the capture of the women angels, Gillmore recreates the primal beginnings of female economic dependence under marriage that was lamented by Gamble and Gilman. For Gillmore, the angels are a 'metaphor for female autonomy' and economic freedom (Rich 2004: 184–5). When the men capture them and clip their wings, Gillmore dramatises the beginning of the institution that still plagued society. The way the men capture them provides a warning for women readers: the men collect mirrors, clothes and jewels from the shipwreck and use them to lure the angels into a 'clubhouse', their materialistic whims overwhelming their good sense. Once they are trapped, the men move in:

> Frank handed each man a pair of shears.
> 'I sharpened them myself,' he said briefly.
> Heads over their shoulders, the girls watched.
> Did intuition shout a warning to them? As with one accord, a long wail arose from them, swelled to despairing volume, ascended to desperate heights. (1988: 196-7)

Feminists such as Gilman had incorporated elements of captivity narrative into stories like 'The Yellow Wall Paper' (1892), with the domestic angel recast as a prisoner trapped behind the bars of the domestic space. In *Angel Island* Gillmore makes this literal, with the angels captured and forced to marry and be dependent upon the men. Though the first half of the book is from the point of view of the men, this forced marriage is represented as a tragedy that appals even the men who perpetrate it. The angels are so injured from the attack on their wings that they barely survive the ordeal. Gillmore turns the Victorian image of the 'angel in the house' into a grotesque and mutilated image of women dominated violently by men who are under the thrall of their evolutionary essences.

Once the angels learn English, the perspective of the novel shifts and the angels begin to narrate the misery of their captivity within the homes built for them by the men. The leader of the angels is a woman named Julia who refuses the name the men try to give her, rejects marriage in spite of her captivity and eventually leads an uprising against the men. Julia is described as an 'Amazon' who works in secret to strengthen her legs and regain the power of flight (1988: 308). However, Gillmore does not cast her as a violent woman who will rise up and kill the men. Like Lane and Gamble, Gillmore separates violence from Julia's nurturing evolutionary essence. Julia embodies all of the ideals of strong independent womanhood without the many flaws – such as resorting to violence – that are characteristic of the men. As Batya Weinbaum argues, Gillmore understood Amazons to be 'women who were taking the lead in situating women to become peers of men, using methods accepted by the current society' (Weinbaum 1999: 13). The revolution Julia leads on Angel Island is non-violent and focused on helping the captured women become stronger to win their freedom of flight back from the men. As such, she represents the ideal feminist that Gillmore later described in her 1931 history of the women's movement entitled *Angels and Amazons*.

Gillmore's feminist uprising looks very different from those of other Darwinian feminists. The non-violent revolution of *Angel Island* is precipitated when a young girl named Angela is born with wings and the men agree to clip them when she turns eighteen. The women, who have never learned how to walk properly since having their wings cut, stumble around helplessly on their underdeveloped feet and are completely dependent upon the men for everything. The first act of rebellion is that Julia inspires the women to force themselves – and the young Angela – to learn to walk as well as the men. This is done covertly, so that the men still believe them to be helpless. Then one day, the women pack up the children and march to a retreat where the men cannot find them. In the first of a series of stand-offs during their 'rebellion', the angels come to the men and Julia informs them,

> we have decided among ourselves that we will not permit you to cut Angela's wings. It means that rather than have you do that, we will leave you, taking our children with us. If you will promise us that you will not cut Angela's wings nor the wings of any child born to us, we in our turn will promise to return to our homes and take our lives up with you just where we left off. (1988: 332)

This stand-off encapsulates the contemporary push of feminists for the right to vote and for economic independence. When the men repeatedly refuse to agree to their demands, the final stand-off ends with the angels taking their children and flying to show that their rebellion has included regaining a limited ability to fly. The men beg the women to return and acquiesce to their demands, with the most sexist of them resorting to 'hysteric entreaty' in his pleas (348). Gillmore inverts not only the power structure of the family – with liberated women asserting control due to their connection to their children – but also places the men in the position of the irrational hysterics who cave in because of their emotions. She also rejects the idea that women are without sexual ardour – something Blackwell, Gamble and Gilman emphasised strongly – and makes clear how much her angels love and desire their men. For Gillmore, this is an important factor to account for in the fight for equality.

The male-female couplings of Gillmore's novel present interesting contradictions that indicate her uneasiness with the whites-only nature of earlier feminist utopias. At the same time, Gillmore's descriptions of her characters hew closely to contemporary ethnological accounts of races and their supposed characteristics, descending at times into crude stereotypes. Julia, the leader of the angels, is described as 'blonde' with 'white wings' who possesses high intelligence (1988: 76). Julia is paired with the blond-haired Billy Fairfax, who refuses to pressure her into marriage even though he has clipped her wings. The scientist leader of the men is Frank Merrill, who 'looked the viking' with his 'blue eyes' and huge size (23). Frank also argues for emancipating the angels, asserting to his male companions that they are 'our equals in every sense – I mean in that they supplement us, as we supplement them' (323). Through these characters, Gillmore maintains the association of northern Europeans with superior intelligence, leadership, sympathy and self-discipline. Frank pairs with Chiquita, who 'looks like a Spanish woman' and is 'lazy' (78, 104). In many ways, she is the opposite of the Frank, but their pairing points toward a future that is interracial as opposed to the belief that pure northern European pairings are essential for the development of an ideal society. Other pairings are clearly supposed to be closely aligned racially. Pete Murphy is described as 'an Irishman' with 'all the perception of the Celt', and he pairs with an angel named Clara who has 'a mop of red hair' (34, 77). Honey Smith is described as 'brown-skinned, brown-eyed, brown-haired, his skin was a smooth as satin' (32). He pairs up with an angel nicknamed 'Lulu' who is 'all copper-browns and crimson-bronzes' and who looks like a 'gypsy, Indian, Kanaka, Chinese, Japanese, Korean – any exotic type that you had not seen. Which is to say that she had the look of the primitive woman and the foreign woman ... Her eyes took a bewildering slant' (101). With such a diverse mix of races on the island, Gillmore makes clear that future generations will be multiracial and that this is a desirable outcome. Though this multiracial vision is limited severely by Gillmore's adherence to racial stereotypes, it provides a stark contrast to the utopian visions of Lane and Gilman.

Within the SF framework of Gillmore's story, the mixing of 'earth men' with the winged angels also constitutes a type of racial interbreeding that runs counter to many eugenics narratives of the early twentieth century. This type of intermarriage provides a final twist in Gillmore's novel that delivers both a utopian hope as well as a harsh reminder of the dangers reproduction holds for women. With the success of her revolt and with her freedom returned, Julia asks Billy to marry her. At the conclusion of the novel, Julia gives birth to a boy and dies soon afterward. Her final words to Billy are, 'My husband – our son – has – wings' (351). While it is tempting to see Julia's death as an example of the frail flower stereotype, her strength and Amazonian bearing throughout the novel instead make her final act seem like a particular kind of feminine heroism, facing danger for her family and the future of this new hybrid race. This ending also shows Gillmore's idiosyncratic take on eugenics: by allowing women to have their freedom, the men in *Angel Island* enable the development of a better race of females and males. The winged baby boy represents what women's liberation offers to men and to their sons. By isolating the characters on an island, Gillmore is able to describe their fusion in terms of natural selection and mutual desire in a manner that avoids state intervention and negative eugenics.

The triumph of scientific femininity: Charlotte Perkins Gilman's *Herland*

Like her feminist predecessors, Charlotte Perkins Gilman used the colonial lost race narrative to present a utopian vision of an Amazonian society in *Herland*. Serialised in Gilman's magazine *Forerunner* in 1915, *Herland* has become the most famous feminist utopia of the suffragette era. The novel also mirrors many aspects of Gillmore's *Angel Island*. The novel is narrated from the perspective of Vandyck Jennings, a trained sociologist, who goes on an expedition with his friends Terry and Jeff. Like Gillmore, Gilman uses her sociologist narrator's evolutionary mindset to provide explanations of sexual selection and its limitations when men dominate it. Vandyck is also the most receptive to the Darwinian feminism of the Herlanders because his rational mind 'cannot deny the civilized progress of Herland' (Hausman 1998: 495). His conversion also fulfils the common utopian trope of the sceptical narrator who learns to love the newly discovered world. Adopting a male narrator allowed Mary Shelley, Gilman and most early

SF women writers 'to enact vicariously a tale of adventure, a triumph of science, in a sexist society that rarely allows the female person such freedoms' (Donawerth 1997: xxiv). Where Gillmore's novel dramatised the horror of domestic captivity for women, *Herland* inverts this with the men becoming captives who are coerced into learning and adhering to the rules of matriarchal culture. While horrible for one of the men, this captivity is represented as an enlightening experience for Vandyck, who narrates point by point the progressive worldview of Darwinian feminism.

Gilman also uses each of her male characters to represent a different model of masculinity, especially as it relates to science and their attitudes toward women. Vandyck notes on the first page that 'All of us were interested in science' (Gilman 1999: 3). Terry is a rich adventurer with expertise in 'mechanics and electricity' as well as 'geography and meteorology' (4). Terry is a clear example of the violent and 'intense masculinity' that Darwin outlined in *Descent of Man* that was championed by Roosevelt and Burroughs and critiqued by Blackwell and Gamble (Gilman 1999: 131). This form of masculinity had also been the focus of much of Gilman's critique in *Women and Economics*. Terry's predatory expectations for women line up with Darwin's model of sexual selection among humans, and he unsuccessfully attempts to capture, possess and 'master' women throughout the novel (Gilman 1999: 131). Jeff is a 'poet, botanist' and medical 'doctor' who is 'full of chivalry and sentiment' and 'idealized women in the best Southern style' (3, 11). He quickly adapts to Herland, but repeatedly notes the Herlanders' lack of traditional femininity. Vandyck 'held a middle ground' between the two others, and notes that he 'used to argue learnedly about the physiological limitations of the sex' (11). Each one of them represents a type of scientific masculinity and a particular manifestation of sexism in American culture. Their captivity and reform – and their many debates along the way – provide Gilman a mechanism for inverting the feminist captivity narrative where women are trapped in the domestic sphere. With men trapped in the large domesticated space of Herland, Gilman shows how certain masculine traits are incompatible with 'civilisation' and a sustainable form of progress.

The sexism of these three scientific men is closely tied to the colonial imagination. Gilman's novel begins with 'a big scientific expedition' where the three men hear of a 'strange country where no men lived – only women and girl children' (4). The expedition is along a river that is clearly supposed to represent the Amazon, and the further they progress up the river, the 'more of these savages had a story about a strange and terrible "Woman Land" in the high distance' (4). The three men decide to mount a second expedition of their own to investigate this rumour, which they consider untrustworthy because it comes from the 'savage dreams' of the local natives (4). The racist dismissal of those seen as lower down the evolutionary hierarchy is compounded by the characters' sexist belief that a society of women could not function without men. Vandyck opines that what they might find is a society 'built on a sort of matriarchal principle' where the men live separately but make 'an annual visit – a sort of wedding call' (9). Jeff argues that it will be a 'peaceful, harmonious sisterhood' that is 'like a nunnery' (10). Terry adheres to the Darwinist man the toolmaker formulation where if women are in charge, 'we mustn't look for inventions and progress' (11). Linking progress with masculine inventiveness was a part of Darwin's argument that Gilman rejected, and she puts that argument in the mouth of the most reprehensible of the colonising men.

Like her contemporaries, however, Gilman accepted Darwin's formulation that women are inherently cooperative by nature due to their maternal instincts. When the three colonial explorers finally 'discover' the Amazonian society, the Amazons themselves are decidedly nonviolent. Though the men provide them with many opportunities for battle, the Herlanders instead choose to subdue and educate them. The only time women express a penchant for violence in Herland is when men try to rape them. Similar to Lane's *Mizora*, the founding moment of Gilman's non-violent utopia was when women killed the few men who remained after a series of wars and a natural catastrophe. As told in Herlander histories, these men were mostly slaves who 'seized their opportunity' and 'killed their remaining masters' with the intention of taking 'possession of the country with the remaining young women and girls' (1999: 56). However, 'instead of submitting', the young women 'rose in sheer desperation and slew their brutal conquerors' (56). Toward the end of the novel, Terry attempts to rape his would-be Herlander spouse Alima, leaving her 'in a cold fury. She wanted him killed' (131). Gilman's Herlanders have 'no fear of men' and 'are not timid in any sense. They are not weak; and they all have strong trained athletic bodies' (130). Terry's attempt at marriage through rape is foiled because 'the sturdy athletic furious' Alima 'master[s] him' physically (140). In this way, Gilman endorses violence as an appropriate response to the violent enslavement of women that Darwin outlines in his vision of sexual selection among humans. In accordance with Gamble's interpretation of the Amazon figure, the women in Gilman's *Herland* are violent only when defending themselves from rape and male attempts to control sexual selection.

While capable of violence, Gilman makes motherhood and cooperation the dominant features of her Amazons. The Herlanders progress 'not because of competition' and wars, but instead 'by united action' (1999: 61). When a miraculous birth through parthenogenesis occurs, they establish 'the Temple of Maaia – their Goddess of Motherhood' and embrace a new method of reproduction without men. They also refuse to make the mistakes of men: when confronted by the Malthusian problem of population in their isolated realm, they did not engage in 'a "struggle for existence" which would result in an everlasting writhing mass of underbred people' (69). They also did not 'start off on predatory excursions to get more land from somebody else' (69). Gilman explicitly rejects this colonial model of progress, instead elevating a model where women control birth and are not trapped in a 'helpless involuntary fecundity' (69). Women take control of their evolutionary destiny through taking control of reproduction, making the warping influences of Darwin's sexual selection irrelevant for their survival. Instead of conquest, Herlander society revolved around children as 'the one center and focus' of their 'thoughts' (67). Like Lane, Gilman's elimination of men does not remove her Amazons from a heterosexual economy of desire: same-sex desire is never presented as a possibility for Herlanders, and the reintroduction of men is seen as a natural moment to reignite some sort of dormant sexual desire. Thus, while she rejects conquest as masculine, Gilman does not reject the idea that sex is only for reproduction and that men are the only natural objects of sexual desire for women.

As the 'natural cooperators', Herlanders also work in harmony with nature to develop a sustainable society (1999: 68). Describing Herland's thoroughly manicured landscape of useful trees and plants, Vandyck says with admiration, 'Here was Mother Earth, bearing fruit. All that they ate was fruit of motherhood, from seed or egg or their product' (61). He also admires their advanced methods of composting and fertilising: 'These careful culturists had worked out a perfect scheme of refeeding the soil with all that came out of it' (80). Echoing the argument of Gamble, Vandyck observes that the sciences of the Herlanders build on their minds, which are 'keen on inference and deduction' (65). Progress is decoupled from Darwin's masculinist framework and the colonial mindset of modern science, and instead focuses on a scientific femininity rooted in maternal cultivation. Herland proves to Vandyck that 'the pressure of life upon the environment develops in the human mind its inventive reactions, regardless of sex' (103). However, this land built on feminist progress is represented as superior to the smoke-filled, poverty-strewn hellscape of American cities. With technologies such as 'electric motors', Gilman's Herlanders point toward a better future through scientific femininity that contrasts sharply to the illness and oppression provided by masculinist science (44).

Probably the most controversial aspect of Gilman's novel is her clear embrace of positive and negative eugenics coupled with her use of racial hierarchies to emphasise the 'civilised' nature of Herland. The feminine science of Herlanders has perfected domestic selection, the foundation for eugenics, which is exemplified by cats that 'were rigorously bred to destroy mice and moles and all such enemies of the food supply; but birds were numerous and safe' (51). Curbing the unnecessarily destructive impulses of cats fits with the overall view of Herlanders minimising violence while maximizing fruitful life. In discussing the advanced nature of the parthenogenetic Herlanders themselves, they explain to Vandyck that their biological progress is due 'partly to the careful education, which followed each slight tendency to differ, and partly to the law of mutation. This they had found in their work with plants, and fully proven in their own case' (78). In this way, Gilman paints a picture of biological diversity without sexual reproduction. However, their progress is also due to 'negative eugenics': those who are 'held unfit' must 'forego motherhood for her country' in its long-term best interests, a sacrifice that is 'appalling' and painful for women whose society is organised around motherhood (70). Vandyck also notes that Herlanders 'were of Aryan stock, and were once in contact with the best civilisation of the old world. They were "white"' (55). This superior 'stock' has been 'toughened somewhat' by the Herlanders' 'heroic struggle' and developed 'unknown powers in the stress of new necessity' (58). The superior nature of their civilisation is magnified by its proximity to the 'savagery' of the nearby natives in 'the dim forests below', another of the many uncritical evocations of racist anthropological hierarchies Gilman makes in the novel (64). Gilman's vision of utopian civilisation is therefore built on a combination of white racial purity, evolutionary struggle and eugenics in a manner that is consistent with the Darwinian feminists who preceded her (Roberts 1993: 75). The notable exception is her contemporary Gillmore, whose *Angel Island* made some flawed attempts to move away from eugenics and white exceptionalism.

The performance of gender is one of the most cuttingly critical aspects of *Herland*. For Gilman, Herlanders represent women who have returned to their natural state of femininity because they no longer have to please men in order to survive. To the eyes of the men, this makes the Herlanders seem downright androgynous. Vandyck notes that Herlanders,

> were strikingly deficient in what we call 'femininity.' This led me very promptly to the conviction that those 'feminine charms' we are so fond of are not feminine at all, but mere reflected masculinity – developed to please us because they had to please us, and in no way essential to the real fulfilment of their great process. (1999: 60)

He later explains that, 'these ultra-women, inheriting only from women, had eliminated not only certain masculine characteristics ... but so much of what we had always thought essentially feminine' (69). Gilman denaturalises supposedly feminine things such as long hair, weak bodies and elaborate, restricting fashions. Gamble saw women's clothing fashions such as corsets as an exemplar of their physical oppression, and Gilman imagines the fashion of liberated women to be 'simple in the extreme, and absolutely comfortable' with an emphasis on function over spectacle (28). When the men put on the clothes of the Herlanders, hyper-masculine Terry grumbles that it contributes to his 'feeling like' they are 'a lot of neuters' (28). Because of their long period without men, the Herlanders have lost the idea of 'sex attraction' as a basis for marriage. Instead they look to other animals such as birds, which are monogamous but 'never mate except in the mating season' for reproductive purposes (124). Here Gilman echoes the argument of Blackwell and Gamble that the unnatural lusts of men were another warped by-product of males controlling sexual selection. She posits that natural gender roles are not polarised, but rather are closely aligned to each other both physically and culturally with only moderate distinctions between the sexes. In the process, she castigates the masculine perspective of Terry as a dangerous product of a stagnant male-dominated society.

Drawing on a convention from domestic fiction, Gilman puts forward Vandyck as an ideal man and romantic interest for the New Woman of the twentieth century (Yaszek and Sharp 2016: 6). The arrival of the men spurs the Herlanders to explore becoming a 'bi-sexual state' again, and Vandyck learns his own fitness for this new type of evolutionary romance (1999: 88). When meeting Herlanders, he finds his 'crowd is the largest' among the three men despite having never been 'popular' among women in the United States (87). When explaining this, a Herlander named Somel notes that Vandyck seems more like a Herlander than the other men in that he seems 'more like People', exhibiting 'characteristics ... which belong to People' and not just to one sex. Gilman redefines the human ('people') based on the Herlanders, who are compassionate and curious instead of violent and colonising. The most desirable males for Herlanders – and by extension, for Darwinian feminist women everywhere – are scientifically minded men with a relatively low libido who model the 'sensitive and sympathetic readers of nature' version of scientific masculinity and who can therefore be persuaded by facts (Milam and Nye 2015: 5). Such men also value the intelligence, strength and independence of women like the Herlanders, making them perfect partners in a progress driven by scientific femininity as much as by scientific masculinity.

Darwinian feminism provided a white educated woman's vision of the future that was predominantly socialist in its politics. That socialism was usually limited by an adherence to contemporary beliefs about race, heterosexuality, ability and civilisation. *Herland* is in many ways the most fully developed exemplar of Darwinian feminism's speculative dimension during the first two decades of the twentieth century. This is partly due to the fact that it was published near the end of a long period of social activism that provided an urgency and focus to Darwinian feminism's intellectual missions. In the United States, women's suffrage became a reality in 1920, and Darwinian feminists shifted more focus to causes such as the birth control movement. The year 1920 saw the publication of birth-control pioneer Margaret Sanger's bestseller *Woman and the New Race*. For Sanger, 'female reproductive autonomy through economic independence and scientific advancement was the only surefire path to healthy families and emancipated women' (Hamlin 2014: 149). Though Gilman did not respond positively to Sanger's call for support, 'American women socialists' and men such as Hugo Gernsback embraced 'the possibility of objective science' and 'reformist readings of evolution' to address the question of sex (Hamlin 2014: 149). Gernsback eventually began to publish the magazine *Sexology* in 1933 to further the cause of examining sexuality through a scientific lens, but by that time his specialist SF magazines had already provided a new banner around which Darwinian feminist speculation could rally. In the pages of Gernsback's magazines such as *Amazing Stories* and *Wonder Stories*, Darwinian feminist visions of the future continued to evolve and challenge masculinist traditions of science and SF.

4 Women with Wings: Feminism, Evolution and the Rise of Magazine Science Fiction

The rise of magazines specialising in SF provided women with new venues for spinning female-friendly visions of the future. Though the beginning of the magazine SF period is often marked as April 1926 – when Hugo Gernsback published the first issue of *Amazing Stories* – SF was being published steadily in magazines such as *The All-Story*, *The Black Cat* and *Weird Tales* in the 1900s, 1910s and 1920s. However, Gernsback provided a name, canon and narrow marketing strategy for the genre, and his magazines proved to be particularly welcoming to women writing SF in the late 1920s and early 1930s (Yaszek and Sharp 2016: xix). As Jane Donawerth demonstrates, women writing SF for magazines such as Gernsback's *Amazing Stories* and *Wonder Stories* tended to imagine developments in 'technologies of childbirth', 'scientific child rearing' and 'domestic duties' that freed women 'for further education and for public responsibilities' (1997: 14–15). These developments in women's SF showed the influence of earlier Darwinian feminists such as Antoinette Brown Blackwell, Eliza Burt Gamble and Charlotte Perkins Gilman. As historian Kimberly A. Hamlin argues, these Darwinian feminists 'grappled with questions of biological sex difference, the extent to which maternity did (and should) define women's lives, the equitable division of household labor, and female reproductive autonomy' (2014: 2). The women of Gernsback-era SF extended the critical vision and speculative dimension of Darwinian feminism, providing possible answers to the problems identified by their predecessors. In this way, Darwinian feminism became one of the unique attributes of SF written by women before the Second World War. At the same time, women SF authors of the 1920s and 1930s began to reframe, qualify and critique Darwinian feminist arguments and assumptions as they wrote them into the genre system of the new SF magazines.

In many ways, the SF written by women during the Gernsback era was similar to that of the men. Like their male counterparts, women SF writers such as Clare Winger Harris, Lilith Lorraine and Leslie F. Stone grappled with evolutionary formulations of colonisation, progress and race that were woven into the genre by influential writers such as H. G. Wells and Edgar Rice Burroughs (Sharp 2007: 70–84, 92–6). However, as outlined in the introduction, there were five frequent storytelling tactics that proved particularly popular among the women SF writers of the 1920s and 1930s that align clearly with the writing and concerns of earlier Darwinian feminists. The women science fictioneers of the period worked within a rapidly changing historical landscape and an emerging constellation of SF sub-genres that also led them to modify, expand and critique the Darwinian feminism of the past. Women SF writers of the Gernsback era generally retained the Darwinian feminist focus on the reorganisation of sexual selection and confirmed the importance of females gaining full control of their bodies. In SF magazines, this often took the form of avoiding a malevolent suitor and finding a man who appreciated a smart and strong woman. The second tactic employed by women SF writers was the fantastic expansion of the domestic sphere. The clearest predecessor for this was Gilman's *Herland* (1915), where a nation run by women domesticated an isolated landscape and built a perfect society free from masculinist oppression. Women SF writers of the 1920s and 1930s often extended this domestic sphere to include spaceships, entire planets or other dimensions. This expansion of the domestic sphere often went hand in hand with a third storytelling tactic, one where women SF writers explicitly gendered and reconfigured the colonial gaze so that women are seen as the carriers of civilisation who must overcome the savagery, arrogance and narrow-mindedness of men. In such stories, traditional SF male heroes were often recast as misguided children or malevolent villains who must be taught a lesson by stronger, smarter females (sometimes from an alien species). A fourth storytelling tactic commonly employed by women authors in early magazine SF was to warn of the dangers of masculinist science, and at times to offer a feminist science as a corrective or alternative. In a departure from earlier Darwinian feminists, the women's SF of the 1920s and 1930s frequently drew inspiration from Mary Shelley's work to dramatise mad male scientists engaging in unethical experiments that negatively impact women. The fifth common storytelling tactic embraced by women SF writers of the

Gernsback era was to represent Amazons and angels as the apex of evolution. Unlike previous generations of Darwinian feminists, however, most women SF writers of the 1920s and 1930s portrayed their Amazons and angels as fully capable of violence and less driven by nurturing motherhood.

Amazons by men

The way Darwinian feminists used the Amazon figure was one of the sharpest distinctions between their writing and the writing of their male contemporaries. In the hands of male writers, Amazons often existed only to be tamed by male heroes. The most progressive image of the Amazon figure in the writing of men of the Gernsback era was the Amazonian flower, a heroic warrior woman who fights alongside her man but who faints in battle or gets captured in a way that turns her into a damsel in distress. The most well-known SF version of this captivity narrative figure is Wilma Deering, who was created by Philip Francis Nowlan in 'Armageddon 2419 A.D.' and its sequel 'The Airlords of Han' that were published in the August 1928 and March 1929 issues of *Amazing Stories*. Wilma is the woman soldier who pairs up with Anthony (later changed to Buck) Rogers in his popular adventures that appeared in magazines, comic strips and movie serials. Like the American Amazons of captivity narratives, Wilma is seen as justified in graphically butchering the racially different males who threaten her. Likewise, Wilma is a feminist whose skills while in soldier drag are celebrated like in the female soldier stories of the nineteenth century. In the race war with the Mongols, Wilma and the other women are not simply love interests or mothers, though their status as such is emphasised. Nowlan depicts women as essential contributors to the war effort, and not simply as curious novelties brought about by exceptional circumstances: they are everyday contributors to progress and the struggle to survive. At the same time, their evolutionary essence at times overcomes them and causes them to become helpless at key moments.

In Nowlan's original stories, the Americans of the twenty-fifth century live a life that is divided between working in factories and serving in the military. Men and women alike rotate through these duties, which are seen as necessary to rebuild American military and technological might to the point where they can attempt to reclaim America from the Mongolian overlords who are concentrated in major cities. When Anthony 'Tony' Rogers awakens from a long radiation-induced slumber in a cave, the first person he sees in the twenty-fifth century is the woman soldier Wilma Deering. She is engaged in a pitched battle with an outlaw gang, and she is outnumbered and alone. At first, Tony's twentieth-century eyes don't see her as a woman: 'I noticed a figure that cautiously backed out of the dense growth across the glade ... The boy's attention (for it seemed to be a lad of fifteen or sixteen) was centred tensely on a heavy growth of trees from which he had just emerged' (Nowlan 1928: 424). This description shows that the immediate gendered judgement of Tony is based on the clothing and combat role being played by Wilma. However, Tony soon figures out Wilma is a woman because she trips, cracks her head and falls unconscious despite a repeated emphasis later in the text on Wilma's superior agility. With Wilma unconscious, Tony picks up her rocket gun and quickly figures out how to use it better than Wilma: he dispatches several of her antagonists, and the remaining enemies retreat. Once Tony has an opportunity to examine Wilma more closely, he begins to emphasise her reassuringly feminine traits: 'Now I had time to give some attention to my companion. She was, I found, a girl, and not a boy. Despite her bulky appearance, due to the peculiar belt strapped around her body high up under the arms, she was very slender, and very pretty' (425). Despite discovering a woman soldier, Tony (and the reader) has the shock of a woman in combat softened by the fact that she is slender, pretty and in need of rescue just like the more traditional damsel in distress. This scenario plays itself out again repeatedly in the stories across media.

This representation of Wilma as both a skilful soldier and a safely feminine woman was consistent with the representation of women during the First World War. As Julie Wosk argues, when women engaged in war work that was traditionally carried out by men during the First World War, representations of these women showed a pervasive fear that they might somehow lose their femininity. These representations compensated for this fear by emphasising the traditionally feminine characteristics of the women to show that even though they were performing the work of a man, they were still 'real' women (Wosk 2001: 191–3). Nowlan's Tony Rogers stories still asserted the essentialist idea that women were physically weaker than men, prone to bursts of irrationality and emotion, and ultimately still needed a man around to save them and lead them into the future. In this way, these stories preserved Darwin's formulation of man the toolmaker: even though women were represented as using tools well, they were still not as adept at fighting or at leadership as men. Women still existed primarily to be fought over and rescued by the men, and were therefore defined primarily through their role in male-dominated sexual selection. At the same time that Nowlan's stories were gaining notoriety, a number of women emerged in Gernsback's publications who provided feminist responses to such masculinist interpretations of Darwin and the Amazon figure. For the next decade, these women became major contributors to specialist SF magazines and the multi-genre magazines that also published SF.

Clare Winger Harris

Clare Winger Harris was the first woman to publish a story in Hugo Gernsback's *Amazing Stories* with 'The Fate of the Poseidonia', which appeared in the June 1927 issue as the third-place winner in a contest. However, that was not her first SF publication: she had already published a story entitled 'A Runaway World' in the July 1926 issue of *Weird Tales*. She went on to author over a dozen SF stories in the late 1920s and early 1930s. As a young woman, she had attended Smith College, which was a leader in educating women in the sciences. In 1886 Smith erected 'the nation's first building dedicated to women's scientific studies and experimentation', and due to demand a new building was erected beginning in 1905 funded by Andrew Carnegie (Hamlin 2014: 63–4). Harris therefore studied science in what were some of the best scientific facilities available for women in the country at the time. By the time she began writing SF, she was a Midwestern housewife with three sons, and she quickly became a popular contributor to Gernsback's magazines (Yaszek and Sharp 2016: 8–9). The range of her stories is impressive, covering such SF staples as mad science, alien invasion, ecological catastrophe, future history, artificial life and forced evolution. Her stories demonstrate a strong familiarity with a wide array of scientific topics, and the discourse of evolution is a constant thread in her work.

'A Runaway World' provided an early vision of global apocalypse that drives people into a Darwinian struggle for survival, something that would become ubiquitous in the SF of the 1950s (Sharp 2007). The coming apocalypse is first understood by an expert on 'atomic energy' and a scientist who has 'established by radio regular communication with Mars, Venus, two of the moons of Jupiter, and ... interstellar space!' (Harris 2011: 17–18). They learn that 'intelligent beings in this vaster cosmos or supra-universe, in which we are but a small molecule, have begun an experiment which is a common one in chemistry, an experiment in which one or two electrons in each atom are torn away, resulting ... in the formation of a new element' (20). Collapsing the cosmic level with the molecular level, Harris shows the possible unintended consequences of scientific experimentation: the 'electrons' Mars and Earth are ripped away from the orbit of the sun and cast into deep space. The story is narrated by a family man named James Griffin, and its primary focus is the impact of science on families and societies as they attempt to survive Earth's long trek toward a new sun. In this way, Harris gives a very personal and domestic perspective on the dangers of science through a vision of 'galactic suburbia' that Judith Merril would apply to the atomic age in the late 1940s (Yaszek 2008: 115–17).

Like earlier Darwinian feminists, Harris takes time to castigate a superstitious religious mindset that refuses to accept the lessons of scientific rationality. When Mars leaves the solar system, the theories of scientists are rejected by many who believe that 'Mars had become so wicked and had come so near to fathoming the Creator's secrets, that it was banished into outer darkness as a punishment' (2011: 21). Harris dismisses this with the scientists' understanding that 'God does not object to His Truth being known', and soon the unprepared anti-scientists pay the ultimate price for failing to heed the scientists' warning (21). Most of the story takes place in domestic spaces where Griffin's family and neighbours gather to watch the catastrophe unfold on television, plan their responses, and go about the process of surviving. Griffin and his family join an astronomer named Marsden, who takes control of the situation for his friends and guides them to his 'observatory' which is 'heated by atomic energy' and can withstand the cold of deep space (27). They gather what provisions they can and transform the observatory into a domestic space split between 'a chemical laboratory', the observatory, and a 'second story' devoted to 'sleeping quarters' (30). As they make the long trek through space, 'the women' show 'more grit and pluck' than the men, laughing and enjoying their time in this combined scientific/domestic space. Bringing together the domestic and the scientific spheres was a common element of Darwinian feminist speculation before the 1920s, and Harris here shows the pleasure that women can have when the two are joined.

When the 'chemical experiment of the super-people of that vaster cosmos' finishes, the survivors on Earth emerge to find themselves orbiting a new sun (38). As they bury their dead and adapt to new conditions, the women prove most 'delighted' with this new situation (38). Though Harris's vision falls well short of some earlier Darwinian feminist speculations in its critique of patriarchal society, she pioneers what would become a common feature of women's writing in SF magazines: women are included as essential and capable partners in both progress and the struggle to survive. Though the women in Harris's stories rarely exhibit scientific expertise – and her stories are completely lacking in strong visions of scientific femininity – she still makes women's perspectives essential to curbing the dangers and surviving the catastrophes caused by masculinist science. She also shows the double-edged nature of scientific progress: while the catastrophe is brought about by a scientific experiment, survival is made possible by the technology of atomic heaters, the scientific rationality of Marsden and the scientific space of the observatory. Even at her most critical, Harris always preserves in her stories a deep commitment to science and the possibilities of progress for women and families.

With 'The Fate of the Poseidonia', Harris reveals a troubling embrace of racial and ethnographic hierarchies common to nineteenth-century evolutionary thought that malingered well into the twentieth century. As Jane Donawerth notes, Harris seemed particularly fearful of the 'mixture of races' and the prospect of white women being forced to breed with non-white men from savage races (2006: 29). 'The Fate of the Poseidonia' is narrated by George Gregory, a man who stumbles across a plot to steal Earth resources by his foreign neighbour Martell. It turns out to his 'chagrin' that his courtship of his girlfriend Margaret is failing due to Martell, who he cannot believe is actually a 'rival' (Harris 2011: 43). He is incredulous because of Martell's 'bodily peculiarities' that include the 'swarthy, coppery hue of his flesh that was not unlike that of an American Indian' (39). Margaret initially chastises George for his intolerance and his jealousy, insisting to him that, 'I refuse to allow you to dictate to me who my associates are to be' (43). Donawerth rightly argues that the 'reader is asked to side with Margaret as she defends Martell and argues for her own independence'

(2006: 29). The story twists, however, as Margaret is duped by Martell. George learns that Martell is from Mars and that his race is stealing water from the Earth along with crafts full of people who subsequently die in the vacuum of space. The story ends with Margaret sacrificing herself by agreeing to stay with Martell's people in an effort to prevent further incursions against the Earth. Harris represents this as a plucky, brave sacrifice that a woman is making for her planet. However, the horror of this sacrifice is based on the stereotype of the menacing, hateful savage who kidnaps a white woman and forces her into bondage. In this captivity narrative, there is no escape and Margaret pays dearly for not fearing other races in the same way as the xenophobic George. Harris would return to this racist trope in her final story 'The Ape Cycle', which included a more traditional vision of the American Amazon.

Harris provided her sharpest critique of masculinist science in the short story 'The Evolutionary Monstrosity', one of a few stories she wrote that centred on bioengineering. The story was published in the winter 1929 issue of *Amazing Stories Quarterly*, and in his introductory headnote, Gernsback referred to her as 'our well known author, Mrs. Harris' (1929: 70). To emphasise the story's scientific basis, Gernsback also published the newspaper article that inspired it, which reported on Dr Ivan E. Wallin's theory of symbioticism (an idea that later influenced the work of Lynn Margulis). In Wallin's theory, bacteria in cells play a role in mutation and changed organisms over time: he concluded that,

> When introduced into living cells these bacteria caused the formation, in many cases, of new tissues and organs, thus pointing to the view that men might have grown up from an original group of cells which developed into a human being through many stages of 'bacterial infection'. ('Wallin's Theory' 1929: 73)

Using this premise, 'The Evolutionary Monstrosity' tells the tale of mad scientist Ted Marston, who discovers a way to harness these bacteria and use them to accelerate evolution without the process of natural selection. Ted transforms his cat into a disgusting creature that walks on two legs and talks, and then begins to use the bacteria to evolve himself, eventually using telepathic powers to enslave his rich friend

Irwin who has been funding his research. In the process, Harris modernises Shelley's critique from *Frankenstein* to show not only the dangers of unmoored scientific masculinity, but also the positive possibilities of bioengineering for women.

The linkage Harris makes between telepathy and evolution was common in SF, and was built upon the link between communication and evolution in *The Descent of Man*: when discussing humanity's evolutionary progress, Darwin attributes human superiority over 'lower animals' to intelligence and language, saying that, 'Through his powers of intellect, articulate language has been evolved; and on this his wonderful advancement has mainly depended' (1998: 49). Linking effective communication to an evolutionary hierarchy – and specifically to the supposed hierarchy between civilised humans, savages and the 'lower animals' – it is not surprising that H. G. Wells gave his highly evolved Martians the ability to communicate telepathically in *War of the Worlds* (1898). Harris follows Wells in this regard, giving a physical description of the evolved Ted that is very similar to Wells's Martians. This description comes when Frank, the heroic scientist and narrator of the story, has a final showdown with Ted:

> Upon a cushion at the far end of the room reposed what looked to me like a phosphorescent tarantula. As I gazed with widened eyes and gaping mouth, I realized that it was not of the spider family at all. The circular, central part was not a body, but rather a head, for from its center glowed two unblinking eyes, and beneath them was the rudiment of a mouth. The appendages which had upon first appearance resembled the legs of the spider, I perceived were fine hair-like tentacles that were continually in motion as if a soft breeze played through them. (Harris 2016: 33)

In essence, Ted has evolved himself down the same dead end as Wells's Martians: the main parts of their anatomies that are preserved are the head and the hands, which are the parts of human anatomy Darwin associated with technology and progress. Ted's head is enormous, his hands have evolved into tentacles, and practically everything else has become rudimentary or vanished altogether. Ted's giant brain provides him a more efficient form of communication, giving him the ability to read the minds of others and impose his thoughts upon them. Harris makes clear, however, that Ted has lost far more than he has gained through his transformation.

Like Wells, Harris provides a critique that does not allow us to simply read this as a type of evolutionary 'hierarchy' that ranges from inferior humans to superior post-human phosphorescent tarantula. Harris represents Ted as a monstrosity, and makes this explicit when Frank says that, 'the evolutionary processes minus the modification of environmental influences point toward retrogression instead of progress. Man dare not tamper with God's plan of a general, slow uplift for all humanity' (2016: 14). While repeating the Spencerian myth that evolution was slow progress, this passage emphasises the Darwinian insight that what is inferior or superior is dependent upon the environment. Harris underscores this with the conclusion to the story: Frank realises that Ted can only control one person at a time, so he engages Ted in a heated debate that allows the enslaved Irwin enough time to shake free of Ted's control, pick up a crowbar and bash Ted into oblivion. In an environment that includes two other men, Ted loses the struggle to survive in spite of his scientifically inflated intellect. Along with showing the folly of Ted's extreme self-transformation, Harris gives a description of his telepathic powers that is rooted in contemporary phobias about radio, a communications technology that allowed the voices and ideas of strangers into intimate domestic spaces. Harris makes this connection explicit in the final showdown: as Ted holds Frank in the grip of his telepathic powers, he gloats that, 'These tentacles are more sensitive than the radio antennas of your era, and they pick up thought waves with little or no difficulty' (24). In other words, Ted has evolved himself into an evil biological radio, developing the means to force his thoughts into the most intimate place of all: the mind of another person. This connection between telepathy, radio and malevolent manipulation became common in the SF pulps, but seems to have been particularly popular among women writers such as Harris (Elliott-Baptiste 2013).

Like Shelley's *Frankenstein*, 'The Evolutionary Monstrosity' provides a critique of scientific masculinity that can be 'read as a warning about the dangers of female exclusion from science' (Roberts 1993: 25). However, Harris also shows that the outcome of this science is not completely negative, holding out the possibility that it could be beneficial to women when in the right hands. The positive possibilities are dramatised through the character Dorothy, the sister of Ted's rich friend: Ted falls in love with Dorothy and attempts to force her into an unwanted marriage. When Frank first meets Dorothy he doesn't like her at all, saying,

> My first impression of her was that she was beautifully and expensively clothed, and I am not a man who ordinarily observes clothes before people. In this particular instance, however, the clothes really possessed more personality than their wearer. The girl was pretty in an insipid, baby-doll way. (2016: 13)

Later in the story Dorothy becomes transformed, showing intelligence and pluck by writing the letter to Frank that warns him about how dangerous Ted has become. When he sees Dorothy again, Frank narrates that,

> I discontinued my ascent of the steps and gazed speechlessly at her, for it seemed I had never seen this girl before – yet I knew it was Dorothy. What refining process had altered her nature and appearance so intrinsically? Trouble is the refiner's fire necessary for some natures, yet somehow this change in Dorothy was not so much one of degree as one of actual difference of quality. (18)

Dorothy explains what happened:

> Ted still loved me, but ... he knew that two beings so far apart in evolutionary development would not be suited to one another, so he intended inoculating me with the germs in order to advance me to his stage of development. Then we two ... would rule the world!' (19)

Because she only receives one dose, Dorothy's change is a positive one. In effect, Harris imagines a cure for frivolous women that makes them more sober and capable of expressing the limitations of masculinist science. Though Frank is the scientific hero of the story, Dorothy clearly puts herself in danger to thwart Ted's plans to conquer the world, thus showing the important role that intelligent women can play in curbing the colonial impulses of masculinist science.

Though Harris does not develop Dorothy into a full manifestation of scientific femininity, she does show Frank as an ideal version of scientific masculinity that is consistent with some aspects of Darwinian feminism. In contrast to Ted, Frank practices science within an institutional framework at his university, and his scientific impulses are guided by morality and a sense of social responsibility instead of being given over to selfish urges. Perhaps more importantly, Harris provides a Darwinian feminist spin on a common feature from nineteenth-century women's domestic fiction: Frank is represented as a sympathetic man who appreciates women of intelligence and strength, who disdains pretty nitwits and who sympathises with the unique struggles faced by women in a man's world. In saving Dorothy from a forced marriage to a predatory tarantula with dreams of world conquest, Frank breaks a cycle of oppression brought about by male control of sexual selection. Dorothy is free to pick her mate instead of being forced to mate with an evolutionary monstrosity. This reproductive emancipation is still tinged with Harris's fear of racial mixing, but in this story at least the primary concern is that women's morality and freedom are not compromised by self-serving masculine violence.

In 'The Ape Cycle', her final published story from the spring 1930 issue of Gernsback's *Science Wonder Quarterly*, Harris added another take on the lonely male scientist who is initially isolated from the tempering influence of femininity. The story focuses on a few generations of male scientists in the Stoddart family who are trying to breed superior forms of 'lower' primates to a near-human level. The evolutionary plot resembles Wells's *Island of Doctor Moreau* (1896), where the dangers and follies of such scientific manipulation without the moral guidance of women are made abundantly clear on repeated occasions. The scientists who dream up this idea are Daniel Stoddart and his son Ray, who are struggling to keep their farm afloat without Daniel's late wife Stella. The difficulty of their life leads to a utopian dream that 'to men and women rightly belong freedom from eternal toil' (Harris 2011: 240). He considers a few alternatives, but explicitly rejects human slavery, noting that the 'enslavement of the blacks had been such an attempt to free the white man from the drudgeries of existence, yet at what a fearful price!' (241). He also notes that the 'age of machines' was not a full liberation because of the fact that there 'will always have to be men to tend machines, and do many other menial tasks' (241–2). Daniel eventually becomes 'more and more enthusiastic over the future possibilities of ape-slavery', despite the warning of his friend Job that 'an innate treachery would prevent these animals from becoming servants in a civilized country' (243). The explicit evocation of simian slavery anticipated the *Planet of the Apes* books and movies of the 1960s and 1970s. Unfortunately, 'The Ape Cycle' has an even more pronounced use of problematic Darwinian anthropology than 'The Fate of the Poseidonia', and Harris's extensive use of racial stereotypes undercuts its putative anti-slavery message. In particular, the simian servants of the story are a SF version of the long-standing 'slave-figure' stereotype, where the 'dependable' and 'childlike' slave of a different race is also 'unreliable, unpredictable and undependable' as well as 'plotting in a treacherous way' (Hall 1995: 21).

Harris reveals the treachery of the simian slaves after an ellipsis of a dozen years, when Daniel and Ray's success has led to a farm cultivated by their improved primates. The landscape is described in terms that are consistent with the Darwinian feminist utopias such as Gilman's *Herland*: 'The two men gazed across the broad fertile acres of the Stoddart farm which were secluded from the outside by an arboreal wall of closely-planted poplars. Everywhere was visible evidence of the scientific care necessary for a perfect cooperation with natural law' (2011: 245). Harris here provides another mixture of warning and progress: on the one hand, they work in harmony with nature, but without a woman around the men have become socially isolated. In spite of their desire to help all humanity, they have failed to see the effects that social isolation has had upon their moral faculties as well as their evolved servants. The biggest danger is made apparent with Harris's description of 'a gorilla' mowing the lawn, a creature described as 'a travesty on the human form' (245). Two gorilla 'overseers' manage an entire staff including baboons and monkeys who turn the Stoddart family's life of hard work into one of plantation-style leisure (246). The arrival of Job's daughter Melva breaks the superficially placid veneer of tranquillity at the house. Ray quickly falls for Melva, but she responds to the idea of evolving apes to be human-like as 'a sacrilege', chastising Ray that 'if they ever do become human we can't have them work for us any more. That would be slavery' (249). Ray responds that with 'the first appearance of a soul ... we will pay them wages and they will be satisfied' (250). This is the first of many instances in the story where women try to keep men on track morally by reminding them of the consequences of their science. At the same time, Harris affirms that women are not anti-science, but rather guides for the best ways to progress. As such, Harris makes her women into partners in scientific progress who curb the worst aspects of scientific masculinity. Harris also has the arrival of women precipitate the treachery of the plantation's simian servants. When Melva meets Alpha, the talking gorilla overseer of Ray's plantation, she is repulsed by 'its extreme unattractiveness' and thinks of it distastefully as a 'missing link' (250). For his part, Alpha 'was gazing fixedly at Melva, and there was that in his too-intelligent bestial face that struck terror into the hearts of ... women' (252). They soon discover that Alpha has murdered Daniel Stoddart, and when Melva is strolling along with Ray, he hears 'a sudden shriek' and turns to see that Melva 'was being born away, apparently unconscious, in the arms of Alpha' (255). This scene is reminiscent of Burroughs's first *Tarzan* novel, where a male from Tarzan's tribe of highly intelligent apes attempts to abduct Jane in a primitive example of marriage by capture. In Burroughs's stories, such scenes become occasions for his white male hero to show his evolutionary superiority over nature and the males of other races by fighting and killing the would-be captor. Rather than repeat this part of Burroughs's evolutionary narrative, Harris inserts a twist of Darwinian feminist irony: Alpha stupidly takes Melva to his home where his 'irate female gorilla' mate Omega rips his throat out in a fit of jealousy and anger over his attempt at infidelity (257). Omega listens to the 'moral instruction' from her 'master' Ray, thus ending the threat of the ape uprising and showing the moral superiority of females in all species. In this way, Harris repeats the common Darwinian feminist claim that females are morally superior to violent, selfish males.

With this initial conflict resolved, the text jumps ahead three centuries and with Melva's help, Ray's success has transformed the world. However, the dangers of this kind of forced evolution persist. Using 'extracts of the known glands of human beings', their son succeeds in breeding specialised primates to be 'gardeners, domestic servants, chauffeurs, mechanics', and every other class of menial help (259). In this way, Harris imagines that this form of progress can lead to the elimination of domestic labour for women as well as physical labour for men. However, they continue to breed more intelligent simians to serve as overseers and to tend the machines so that across the planet, more humans are 'emancipated' from work of all kinds and are free to pursue nothing but leisure. In this future, women remain the voice of moral conscience and warning: Wilhoit Stoddart's mother scolds him that, 'Unless a man is mentally ready for emancipation, he deteriorates instead of progressing' (259). Harris evokes here a standard early twentieth-century argument about overcivilisation, a Darwinist formulation of weakness that was believed to afflict those who no longer struggled to survive and gave in to the decadence and easy living of modern life. Overcivilisation was understood as especially dangerous for men, feminising them in ways that novelists such as Jack London and politicians such as Theodore Roosevelt believed made them unfit for perpetuating the nation, the species and the spirit of the United States' intrepid pioneers (Den Tandt 1996: 641; Sharp 2007: 57). In Harris's story, women understand and speak for nature, warning about the dangers and limitations of ignoring evolutionary laws and processes. In this case, Wilhoit's mother is also the voice of normalised gender roles, as she warns, 'not all people can stand prosperity and the leisure that accompanies it. Take Hayes Sulter for example' (259). Hayes walks in moments later, his 'sneering voice' and 'indolent face' marking him as the embodiment of overcivilisation that makes humanity ripe for overthrow by the intelligent simian slaves. In an ensuing conversation about the status of ape slaves, Hayes argues that ideas about emancipation come from a 'mistaken sense of philanthropy' (260). Here Harris uses language similar to pro-slavery scientist Josiah Nott, who in his antebellum work *Types of Mankind* argued that different races constituted distinct species, and therefore 'no philanthropy, no legislation, no missionary labors' could positively elevate inherently 'inferior types of mankind' (1854: 79–80). Instead of embracing such a static polygenist vision of natural racial hierarchies, Harris puts them into the mouth of a villain to contrast with the heroic, evolution-minded Wilhoit and the optimistic possibilities provided by Darwinian thought. Harris thus rejects the most extreme kind of racist argument that remained common in many parts of the United States during the early twentieth century.

Harris dramatises the relative fitness of Hayes and Wilhoit to survive when the inevitable global simian rebellion occurs. Hayes is quickly dispatched without a fight. Wilhoit, who has never stopped working, is captured while resisting and preserved because his superior mind and knowledge of evolution serves the goals of the simian rebels. This demonstrates the folly of overcivilised manhood, and underscores the fighting chances of those men who embrace a life of work. As expressed by one of the rebels, the plan of the overseers and more intelligent simians is to 'get girls' from the human population for breeding a 'future race' (2011: 274). By doing this, Harris mobilises yet again the threat of racially other males – in this case evolved simians instead of African Americans or Indians – who are biologically driven to rape white women. This racist trope allows Harris to draw upon the colonial figure of the American Amazon. As discussed in chapter three, the colonial logic of the American Amazon made violence by women acceptable when male relatives were not around and they needed to protect their children or defend themselves against rape. Harris's Amazon character is Sylvia Danforth, a neighbour from a nearby plantation who shows her evolutionary fitness soon after her introduction by selecting Wilhoit as her proper suitor while rejecting the advances of the overcivilised Hayes. When the gorilla overseer on her farm attempts to capture her, she guns him down and thus 'escaped a worse fate than death' (279). She then rescues Wilhoit, hands him a gun, and tells him, 'do your duty' (279). Wilhoit finishes off his own gorilla overseer, who is 'a hairy prostrate form' on the ground because Sylvia has already wounded him (279). Like her Darwinian feminist predecessors, Harris rejects male control of sexual selection and its attendant 'primitive' practice of marriage by capture. She also follows the hegemonic racial discourse of her era by having her strong Amazon woman remind her man of his evolutionary duty: to kill racially other males who threaten to rape her and to put them back in their proper social place. Harris preserves the essentialist idea that women are inherently non-violent, as Sylvia only engages in violence long enough to free Wilhoit. Once free and spurred on by Sylvia's words, Wilhoit takes over his proper evolutionary role as her violent protector.

Harris does represent this future as one where women have become relatively emancipated and have proven their abilities in other areas traditionally reserved for men. While Wilhoit makes other preparations, Sylvia repairs the airplane they will need to bomb the stronghold of the gathered rebel leadership. When Wilhoit expresses surprise at her technical skill, Sylvia counters that it 'was purely a matter of environment ... You know women finally came into professions that had been hitherto considered solely man's field, and they found they could do as well as their brothers' (281). In their successful raid on the 'monkey republic' capital, Sylvia takes over as pilot and 'under the deft control of the girl, the machine climbed rapidly upward into the clouds' to escape the detonation of their payload (289). This leads Wilhoit to gush, 'Gee, you sure are some pilot!' (289). In this way, Harris rejects the essentialist arguments of scientific masculinity that assert women are incapable of engaging in techno-scientific pursuits. She also creates one of the first Amelia Earhart figures in magazine SF. Earhart was the first woman to make a flight across the Atlantic Ocean as part of a crew in 1928. Three women and thirteen men had died in the year prior to the success of Earhart and her fellow crewmen, and her success made headlines throughout the English-speaking world. Earhart always represented her many feats of aviation as 'evidence of women's capabilities in the modern world and as a spur to further advancement' (Ware 1993: 42–4). In the period when women first became major contributors to SF specialist magazines, Earhart was easily the most prominent feminist icon associated with techno-scientific progress, so it is not surprising that the 'angels' of Darwinian feminists during this period took on the form of Earhart-like aviators who master the skies with skill that exceeds those of their male counterparts. This also dovetailed nicely with Gernsback's long-standing love of aviation and interest in female aviators that culminated in his creation of the magazines *Air Wonder Stories* in 1929 and *Aviation Mechanics* in 1930 (Ashley 2004: 161–5). In Harris's case, her final story provided her strongest feminist character, a woman who embodies a type of scientific femininity that enables her to work alongside her chosen mate and guide him in the pursuit of techno-scientific progress. Unfortunately, Harris did not jettison the racial baggage that accompanied the American Amazon figure, building the heroism of her Amazon on the rampant racial stereotypes of the day.

Minna Irving

Like the first wave of Darwinist feminists, women publishing in Gernsback's magazines tended to follow Darwin's view of progress as contingent and not inevitable, with many openly advocating for feminist positions in relation to reproduction, domestic labour and women in the workplace. They also tended to see males as potential predators while representing the ideal male as a scientific man who embraced strong, intelligent women without trying to subjugate them. Minna Irving's 'The Moon Woman', which was published in the November 1929 issue of

Amazing Stories, provides another exemplar of this Darwinian feminist influence in early magazine SF. The November 1929 issue was retired natural science professor T. O'Conor Sloane's first as editor of *Amazing Stories*, and it was Miriam Bourne's first issue as managing editor. Both of them had received promotions after founder Hugo Gernsback lost control of the magazine, and their publication of Irving's story indicates that Gernback's practice of publishing feminist stories continued after his ouster. The editorial blurb for Irving's story read,

> MOST of our authors, thus far, have been more or less pessimistic of the future. Just why this should be so, we cannot say. Our new author, however, shows nothing of that fear. Rather, she sees a considerable amount of improvement several thousand years hence. Even effectual communication with another planet does not phase her. If the problem of indefinite suspended animation could be solved, we wonder how many people would lend themselves to such an experiment, even with all chances in their favor.
>
> We are sure you will agree with us when we say 'The Moon Woman' is a beautiful story. (Irving 1929: 746)

Minna Irving was the pen name of Minnie Odell, and while she was new to *Amazing Stories*, she was a woman with a long publishing record who 'saw the science fiction magazines as a new venue' in which to publish (Davin 2006: 119, 232). She was primarily a poet, and she went on to publish a number of gothic and fantasy poems in magazines such as *Weird Tales* and *Famous Fantastic Mysteries*. With 'The Moon Woman', she provided her only SF story published in a specialist SF magazine.

Irving's story is characterised by unapologetic and explicit feminism that tweaks scientific masculinity and provides another flying heroine in the Amelia Earhart mould. The scientist of the story is Professor James Holloway Hicks, who 'discovered the wonderful serum of suspended animation' (Irving 1929: 746). After the machinations of his unscrupulous assistant leave the professor in a Rip Van Winkle kind of slumber, he awakens 200 years in the future in the ornate 'marble mausoleum' he built for his experiment (750). Professor Hicks awakens when 'a young woman' named Rosaria with 'shoulder-bladed extended broad wings of a glittering, semi-transparent, membraneous material' investigates his crumbing mausoleum (751). These artificial wings fold 'by touching a small protuberance set in a belt of white leather that crossed her full bosom' (751). The woman in Irving's story is like the angels in Gillmore's *Angel Island*: she is a modified, feminist version of the angel in the house, representing both the inherent virtue of women as well as her freedom from the house (Donawerth 1997: 49). Women are freed from the home in Irving's story because of the invention of concentrated food. As Rosaria explains, 'We have thus eliminated a great deal of unnecessary work ... The kitchen range and sink have disappeared with the butler's pantry and the storeroom' (753). With domestic drudgery behind her, Irving's winged woman can roam the skies and engage in archaeological discoveries such as excavating Professor Hicks in his mausoleum. However, this freedom is tempered by the threat of predatory males. Rosaria carries a weapon called a 'radiomatic', and she explains that, 'All women carry them ... for since everybody flies who can afford to buy, borrow, rent or steal a pair of wings, it is not safe for any woman to fly out alone without being able to protect herself' (753). Irving's winged women don't need men to protect them – making them a kind of Amazon – and their mastery of weaponry shields them from the Darwinian primal scene where men violently enslave women.

In regard to evolutionary romance, Irving frustrates the usual Edgar Rice Burroughs model where a civilised woman swoons into the arms of the ideal Darwinian man. Instead, it is Professor Hicks who is transformed from a cold man of science into an ardent lover by this ideal model of Darwinian feminism. While Rosaria is climbing back through a hole in the roof of the mausoleum after getting some food for the weakened professor, she drops down and the professor has to catch her:

> With that white and gold bundle of womanhood in his arms, the professor suddenly felt how silly all his crucibles and retorts and serums had been. He could not even remember the formula of the serum of suspended animation, and he didn't care if he never remembered it now; it had served its glorious purpose, it had bridged the centuries between him and this super-girl, who was winged like an angel, and he felt that he was through with all that had been so important to him two hundred years ago.

> For the first time in his two hundred and thirty-five years, the professor was in love. (753)

While estranging Darwinian romance in this passage – with the man transforming instead of the woman – Irving maintains a heteronormative model of romance that some other Darwinian feminists troubled (most obviously C. L. Moore in 'Shambleau'). That said, this passage also shows what the ideal man was for this heterosexual feminist future: he admired the free, powerful Rosaria and wanted to join her in her adventures instead of being driven to violently enslave her and keep her in the domestic realm. In effect, Irving shows how men would have to evolve to appreciate the feminist angel of the future.

Like Gillmore's *Angel Island*, Irving's eugenic vision of the future is interracial: when the professor asks how this utopia came about, Rosaria explains that it was due to space travel. When the first trip of Earth men to the moon occurred in 1930, she says that the people of Earth met with 'the moon-people', who then,

> came to earth ... and finding earth so very far behind moon-times ... the moon-people remained here, and sent for many more. Being so much wiser and so much farther advanced in civilization than the earth-people, they became rulers here, and by intermarriage soon improved the earth-races – mentally, morally and physically. (754)

Irving here inverts the colonial frontier narrative of colonisation so common to SF of the period, with the enlightened aliens colonising humans instead of the other way around. She also harkens back to the older model of utopian writing from women such as Margaret Cavendish and men such as Francis Godwin in the 1600s, where other worlds provide a model of civilisation that makes our own look downright savage by comparison. Rosaria goes on to reveal that she has an Earth father and a Moon mother, and the professor – who was 'hitherto callous to female charms' – falls to his knees and asks her to marry him (754). In contrast to Harris, Irving represents the eugenic improvements from mixing races as an ideal that inspires the white male scientist to abject devotion, estranging the common racist fear that racial mixing would lead to degeneration instead of progress. It also emphasises the importance of alien feminine influence, with the

Moon mother of Rosaria a civilising influence on the savage patriarchy of Earth.

M. F. Rupert

While women published SF in a number of magazines during the late 1920s and early 1930s, Darwinian feminism flourished most in the magazines Gernsback established after he lost control of *Amazing Stories* to bankruptcy in early 1929. At the beginning of the bankruptcy proceedings, Gernsback sent a letter to all of his authors and notified them that he would be establishing new magazines that would pay the same rates (Ashley 2004: 133–5). By the middle of 1929 Gernsback had launched *Science Wonder Stories*, *Science Wonder Quarterly* and *Air Wonder Stories*, and brought with him authors such as Leslie F. Stone (Ashley 2004: 138–9). After a year, he consolidated these titles under the *Wonder Stories* banner, which he published as editor-in-chief until he left the SF magazine business in mid-1936. He also began a mail-order booklet series in 1929 entitled *Science Fiction Series* that published novellas by Stone, Lilith Lorraine and Amelia Reynolds Long. To help with these new publishing ventures Gernsback hired MIT graduate David Lasser as his literary editor. When he consolidated his magazines under the *Wonder Stories* banner, Gernsback promoted Lasser to managing editor. Lasser was a socialist activist who eventually left Gernsback's magazines in late 1933 to lead the 'Workers Alliance of America, a nationwide agitational organization which became the major representative of the unemployed and of WPA workers during the Great Depression' (Davin 1999: 31, 38). Lasser was serious about his socialism and clearly embraced feminism as well: during his tenure at Gernsback's magazines there was a proliferation of positive feminist visions of the future. Unlike Gernsback and T. O'Conor Sloane, Lasser worked with authors 'to shape and mold the field' by 'suggesting ideas, commenting on drafts, and even collaborating' (Davin 1999: 41). As discussed in the previous two chapters, Darwinian feminism was most appealing to social reformers and socialists such as Eliza Burt Gamble and Mary Bradley Lane, so it makes sense that Lasser actively solicited and shaped stories by women authors that closely hewed to Darwinian feminist themes and principles. His tenure with

Gernsback's publications from 1929 to 1933 constitute the golden age of Darwinian feminism in magazine SF.

One story type that Lasser seemed to promote was the 'battle-of-the-sexes' showdown between feminists and entrenched patriarchal cultures. Justine Larbalestier argues that many battle-of-the-sexes stories in early SF sought to 'contain the threat of change to the relationships between men and women', but at the same time made 'explicit the ways in which romantic heterosexual love and sexual intercourse are crucial to the shaping of "real" women's subjectivities' and their dependence upon 'an economy of compulsory heterosexuality' (2002: 11, 13). Battle-of-the-sexes stories in the Darwinian feminist mode engaged in major renegotiations of this economy, at times even rejecting it completely. One clear example of this was printed in the spring 1930 issue of *Science Wonder Quarterly*. M. F. Rupert's 'Via the Hewitt Ray' put forward a Darwinian feminist utopian vision to rival Gilman's 1915 masterpiece *Herland*. 'Via the Hewitt Ray' focuses on an Amelia Earhart figure – an adventurous woman pilot named Marion Hewitt – who has to rescue her clueless scientist father from an experiment that has transported him to the fourth dimension. When she arrives in the fourth dimension, Marion finds herself in the middle of a war between the 'Second evolutionary plane' of the fourth dimension, which is a feminist utopia occupied by Amazons, and the 'Third evolutionary plane' dominated by evil male super-scientists. As Marion gets a tour of the world of the Amazonian 'Seconds', Rupert provides a warning to her male readers through an account of the utopia's history. Mavia, the leader of the Seconds, tells Marion that, 'the men were the ruling sex of this plane, but gradually the women demanded equal rights and once we gained a footing, it wasn't long before we were ruling the men' (Rupert 1930: 377). Mavia notes that, during 'The Sex War Epoch ... the women won, and we destroyed millions of the despised masculine sex' (377). The few men who remain are assigned to jobs as breeders or prostitutes, and have no freedom whatsoever. This female extermination and enslavement of males serves as a warning about the dangers of female oppression, underscoring to male readers that they need to treat women as equals, or else. Unlike the Amazons imagined by Gamble and Gilman, Rupert's Amazons reluctantly embrace violence and engage wholeheartedly in a war of extermination against the men who threaten them. Where Gilman's Amazons engaged in a war of extermination against males, she only mentioned it in passing. Rupert devotes a sizable chunk of her story to the battles her Amazons fight against their male antagonists, taking time to show the zeal women show for violence.

The story stages one final battle of the sexes near the end, this time between the Seconds and the 'horrible grotesque' men of the 'Third evolutionary plane' who have 'tiny weak bodies and enormous heads', along with 'Clever machines [that] carry them around and do the physical acts that their little wizened bodies are incapable of performing' (378). The description of the Thirds hearkens back to the super-scientist Martians of H. G. Wells's *War of the Worlds*, thus marking them as an evolutionary dead end because of their overreliance upon technology. Rupert also extends the critique of science by predecessors such as Mary Shelley: she represents scientific masculinity as a direct threat to women, and shows how male models of progress lead to physical deterioration. To drive home this point, Marion joins the Seconds in completely exterminating the Thirds. In the heat of the battle, Marion even narrates that, 'They were such inhuman-looking creatures – that when they crumbled and disappeared as my ray touched them, I felt no revulsion as I might have had they been more human-looking' (383). In effect, Marion's lack of remorse emphasises how undesirable such an evolutionary future is to women, and points instead toward a more feminised form of progress as the ideal.

The scientific femininity of Rupert's story is among the most developed in pulp SF of the period. Where the male Thirds work to replace their bodies with machines, the Amazonian inventions of the Seconds are focused primarily on community and health. In the spirit of Amelia Earhart, the Seconds are aviators who have fleets of 'enclosed cabin aircraft' in each city to facilitate travel between their cities and defend themselves from hostile patriarchies (375). When Marion first meets Mavia, she is surprised that they can communicate. Mavia explains that, 'By means of sensitized plates within these caps your spoken thoughts' are transmitted and 'translated by the wires on my cap and come to me as spoken in my own language' (377). In effect, the Seconds have developed a technology of telepathic communication that also serves as a universal translator. After women took over the second evolutionary plane, they used the communicators to unite all of their cities and end the constant 'warring upon each other' (377). In her tour of Mavia's city, Marion also witnesses a woman wounded in battle who has had her damaged 'flesh and blood heart' replaced with an 'artificial rubber heart' powered by 'stored up electrical energy' (379). Through these technologies, Rupert paints a picture of progress devoted to uniting people and healing bodies. This scientific femininity is so threatening to the patriarchal culture of the Thirds that they repeatedly attack, forcing the Amazonian Seconds to also invent ray guns. Like earlier Darwinian feminists, Rupert portrays women as essentially cooperative and as the natural builders of long-lasting civilisation. However, she also dramatises their propensity for extreme violence and warfare in a way that belies earlier Darwinian feminist formulations of the Amazons. Though their violence is primarily a result of threats posed by violent men, the women in Rupert's story display a passion for battle and remorseless killing that shows it is also a part of their evolutionary essence.

Rupert also provides a subtle critique of utopias such as Gilman's *Herland* through her account of the eugenics and reproduction of the Seconds. Where Gilman's Herlanders developed a form of parthenogenesis, Rupert rejects that as unscientific nonsense. Marion asks Mavia,

> 'I should think ... with your advanced knowledge of science you would have been able to produce young without the actual help of the male.'
> Mavia laughed heartily. 'We did try it and you should have seen the results. Perfect monstrosities. We did not want our race to deteriorate, so we went back to the age-old method.' (378)

Here Rupert rejects not only Gilman's parthenogenesis but also Frankenstein-like tampering with reproduction: the women are in control of sexual selection on the second evolutionary plane, and it is represented as natural that they breed with those men they find to be desirable. However, the Seconds have also developed eugenic practices whereby the males are 'put through rigid physical and mental tests' and those with a 'high average' are turned into 'reproducing males' (378). Through eugenics, Mavia touts how the Seconds eliminated criminal behaviour and perfected their giant Amazonian bodies. Rupert shows some unease about such feminist arguments for eugenics through the scepticism of Marion. From Marion's perspective, the Seconds have lost all emotion that makes life worth living, including parental affection and romantic love. Their children are joyless and raised by trained insects, and their men are pathetic slaves. Through this speculative inversion of Earth's heterosexual economy, Rupert draws attention to the plight of women in her own society while maintaining the Darwinian feminist insistence on motherhood as the central civilising force of evolution. In the process, she rejects the idea that one sex should rule alone, instead pointing toward a model of gender complementarity similar to the evolutionary arguments of Antoinette Brown Blackwell.

Rupert provides a complementary model of the heterosexual economy through Marion, who is represented as a red-blooded woman who responds viscerally to a rebellious man named Joburza. Marion sees him put on trial for advocating men's rights, and says, 'I suppose, being a woman, it was natural that I should notice his good looks first of all' (380). She then goes on to describe why the men of Earth are unattractive to her: 'I knew I would marry, but somehow or other the men with whom I came into contact either left me cold or, if they did appeal to me, they usually aroused my antagonism by their airs of superiority' (380). After having the rebellious man's life placed in her hands, Marion returns with him and her father to her own dimension. She renames him John and he proves to be the perfect man: he fully accepts her talents and abilities, and he is even grateful to be treated like an equal by a woman. Rupert uses her battle-of-the-sexes story to show that the ideal man is one who respects women and sees them as equals. She makes clear that the dangers for men of treating women like inferiors include extermination, slavery and exclusion from the heterosexual economy. In this way, Darwin's vision of evolution provided Rupert the means to investigate the compulsory heterosexual economy: her version of the battle of the sexes shows how women can take their evolutionary destiny into their own hands and demand equality and social justice, while warning about the dangers of over-reliance upon eugenics and scientific methods of improving the species. Unfortunately, this is Rupert's only SF publication that has been rediscovered to date.

Leslie F. Stone

Leslie F. Stone was the pen name for Leslie Francis Silberberg, a woman who wrote several stories in the late 1920s and 1930s that became increasingly feminist once she began publishing stories with Lasser for Gernsback's magazines. Stone studied journalism in college and married a labour journalist in 1927. She had been writing SF since she was a young girl, and published over twenty stories between 1929 and the mid-1940s (Yaszek and Sharp 2016: 26-7). In a speech at a convention in 1974, Stone said that growing up she read 'Argosy magazine' and also 'discovered Edgar Rice Burroughs, notably *A Princess of Mars*' (Stone 1997: 102). Stone admired Burroughs's writing where 'adventure was presented for the sake of adventure on other worlds and our own' (1997: 103). In many of her stories, Stone was clearly revising Burroughs's 'planetary romance' style and reimagining it from a feminine perspective with stronger female characters and cultures (Attebery 2006: 55). Where Burroughs chose Mars as the ideal planet for masculine adventure, Stone has her heroines zoom past Mars to more interesting planets or populate Venus with a superior matriarchal culture. In the process, she reworked the planetary romance into a far more progressive form that made women the heroes of evolution and called into question many of the colonial conceits woven into the genre.

'Out of the Void' was the first of Stone's stories in this vein to be accepted for publication, though it was the second to appear in print (Stone 1997: 100). Published in the August and September 1929 issues of *Amazing Stories*, Stone drew upon the trope of the woman soldier in drag to insert a woman's perspective into space adventures. Her intervention was far more than simply writing a 'woman's interplanetary romance on a shopgirl level' as Bleiler asserts (1998: 414). Stone's story denaturalises scientific masculinity and adventure heroics, demonstrating how they are gendered performances that women can engage in as convincingly as men. Stone's protagonist, Dana Gleason, had received 'the finest education that man could acquire' before trotting around the globe on adventures with 'his' father and becoming a distinguished pilot 'in the Air Force' during the First World War (Stone 2016: 36-7). As a pilot, she is another example of the Amelia Earhart figure who is in full Amazon mode as a warrior. As the story unfolds, an alien ship returning to Earth brings news of Gleason's successful trip into outer space as the first astronaut. For most of the story's first instalment, Gleason's bravery and strength are noted as other characters try to sort out the mystery of 'his' trip into space in the ship designed by reclusive scientist Professor Rollins. Toward the end of the first instalment, Gleason's diary reveals to both the characters and the reader that she is a woman. The revelation of Gleason's 'true' sex draws attention to the performative aspects of gender, leading the characters and reader to reassess all Gleason's achievements and the potential inherent in women (Donawerth 1997: 157). In effect, Stone asserts that with male clothing and male privilege, women are as capable as men at being warriors, scientists and adventurers.

For the rest of the first instalment and much of the second, 'Out of the Void' is narrated by Gleason via her diary. Stone was therefore among the first authors to use a female narrator in a SF story, a device that allows her to stress the significance of women's abilities. As Gleason defiantly declares in her diary, 'haven't I proved myself equal to any man?' (2016: 53). However, the story's second instalment lapses at moments into a Darwinian essentialism akin to Burroughs in his John Carter stories. This happens through the introduction of a love interest named Richard Dorr. Before the door of Gleason's ship closes for her trip to Mars, Dorr forces himself in and they set off together. Dorr, who has learned her secret, explains to the surprised Gleason why he came along:

> Oh, yes, I know your records ... I know all your courageous deeds, your researches, your science, your war experiences, your bravery. I know all that, but with it all ... you *are* a woman. You are brave, strong, great willed, yet you *are* at a disadvantage, and you are attempting a tremendous thing. (55-6)

Early in the second instalment of the story, Gleason shows why she is at a disadvantage as she collapses under the physical and emotional stress of the space voyage. After Dorr carries Gleason to the couch and kisses her, he declares, 'How I love you. I loved you before I knew you were a woman, and I have adored you more each day ... I thought that on this other world we should find love together, in the need for each other, work together, live for each other' (61). The initial strength of Gleason's character, and the queerness of Dorr's romantic interest in her as a man, indicates that Stone was testing the waters with subverting hegemonic notions of gender and sexuality. However, once under the spell of love – or in Darwinian terms, sexual selection – Gleason becomes a plucky but swooning beauty who is less suited than a man for Burroughs-like adventures (Donawerth 1997: 157-8).

Despite lapsing into an essentialist romance plot device, Stone continues to introduce subversive feminist elements into the second half of 'Out of the Void'. When Gleason and Dorr zoom past Mars and crash land on an alien world called Abrui, the narrative voice of the story shifts to the male alien Sa Dak who brought back her diary. He proceeds to narrate the remainder of the story, but the focus remains mostly on Gleason's experiences in the royal courts. Dorr is the hero who leads an army to overthrow the repressive government, but most of his actions occur elsewhere and are told to Gleason in brief by other characters. For her part, Gleason 'made it a point to meet women, to learn what she could about them, to understand them. She remembered her promise, made aboard the rocket, to accept her rightful heritage as a woman, and to do what she could to aid her sex' (2016: 81). What she finds on Abrui is that the life of 'woman was not an unhappy one. Women, as well as men, did their part in the nation's work, and wives were as well acquainted with statecraft as were their husbands. Nor were they denied the right to do any manner of work they desired' (82). Stone uses this distant world to imagine a form of gender equality that is in stark contrast to the Earth she left behind, where the only way a woman could get ahead was to pass as a man. Stone also develops a version of scientific femininity through Gleason. Where Dorr uses his knowledge of warfare to help create weapons for the oppressed peoples of Abrui, Gleason tours the planet lecturing to the ruling Tabora on 'Earth, its peoples, its thinkers, its astronomers, [and] its theories' (80). She also uses her mastery of communications technologies to become 'the inventor of the radio by which Tabora could now transmit the voice through the air' (84). The aliens of Stone's Abrui have no problem accepting scientific femininity, and without sexism her heroine is able to be both a woman and a scientist without this seeming to be a contradiction.

While Gleason is exalted as a great explorer and inventor, Stone's 'Out of the Void' provides a harsh critique of colonialism and the use of race to create social hierarchies. Sa Dak, the alien narrator of most of the story's second instalment, describes the 'three races' of Abrui, beginning with the silver-skinned Tabora who 'esteem themselves the only civilized people of the planet, though once they were barbarians occupying the "backlands" into which they drove the Gora, the simple-minded' race that is 'bronze-skinned' (67). The slaves of the Tabora are the golden-skinned Moata. Sa Dak is himself a Tabora, and his description of this racial hierarchy seems to be reinforcing the perspective that Abrui's social hierarchy is a natural outgrowth of racial characteristics. However, at the end of the story it is revealed that

Sa Dak is really Moura, the primary villain of the story who tries to force Gleason into marriage (Donawerth 1997: 159). His narration is one long critique of his own folly and the hubris of his people, a revelation that recasts his racial descriptions as the fantasies of colonisers who were deluding themselves. The degraded conditions of the Gora and Moata are explained as products of colonial processes, not inherent biological traits: Dorr explains to Gleason that they are not 'the barbarians the Tabora make them out to be. They were civilised centuries before the Tabora. But in their struggle for existence they have no time to give to science; hence they are fast deteriorating into savagery' (90). In Darwinian terms, Stone presents races as biologically equal, with the Tabora only gaining ascendancy due to technology and colonial power. In this sense, she uses the estranging setting of another planet to critique the racial and colonial hierarchies created by Euro-American expansion. At the same time, Stone reinforces the nineteenth-century idea that levels of technology determine the relative sophistication of cultures. Stone uses the language of 'savagery' and 'civilisation' uncritically regardless of the narrator, with Richard Dorr drawn as a standard white saviour character who has 'the face of a Viking' and who 'was doing a lot to help the natives of Africa find themselves' before his trip into space (44). He continues this kind of work on Abrui, leading an uprising and slave revolt that puts the Gora and the Moata on equal terms with the Tabora: 'Dorr ... had made warriors of the Gorans, and now he made civilized men of them' (102). In the logic of the story, the subjugated races of Abrui are inherently incapable of throwing off their oppressors, and only with the help of a 'Viking'-like saviour can they be 'made' into civilised men. Thus, while Stone explicitly criticises colonialism and racism, the racist premises built into Dorr's character limit the progressive possibilities of the story.

Stone published two more stories in *Amazing* in the next year. Her vignette entitled 'A Letter of the Twenty-Fourth Century' appeared in the December 1929 issue, and her short story entitled 'Through the Veil' appeared in the May 1930 issue. Both stories played with feminist themes and images that were becoming more central to her writing. As its name implies, 'Letter' is a message from a man from the future named Harry to his friend Joe. Harry had found a stash of old SF magazines and gives Joe a series of gleeful plot summaries – including 'the supremacy of woman and the deterioration of man' (Stone 1929: 860) – before giving an account of the utopian world of the twenty-fourth century. In the list of the many ways they have progressed over their forebears, Harry includes 'how to make child-birth a safe and beautiful function', something Stone would return to in later stories (861). 'Through the Veil' blended fantasy with SF in a manner that was not common for SF specialist magazines of the day, particularly *Amazing* under Sloane. The story follows the adventures of two former college chums, one of whom has invented a 'Z-ray camera' that can take fourth-dimensional pictures of fairies, elves and gnomes (Stone 1930: 175). One of the men recounts actually visiting the fourth dimension, which the elves call 'the Veil', which is an idyllic pastoral world untouched by industrial blight. One fairy goes on an extended tirade about the excesses of 'ape-thing' evolution and scientific conquest, including the killing of 'animals for meat' – where humans 'murder them scientifically' – and the building of 'factories' and 'ugly buildings' that 'belch smoke into the God-given air' (178). The fairy adds, though, that 'your women are getting a little sensible these days' (178). The fairy also gives an account of fairy evolution from butterflies, and reads the mind of the man with benign telepathy. The ecological utopia of Stone's story is similar to the garden-like world of Gilman's *Herland*, and is associated with a rejection of scientific masculinity along with the colonising tendencies of Euro-American science.

'Men with Wings' was the second story Stone sold, but it was the first to see print when it appeared in the first issue of Gernsback's new *Air Wonder Stories* in July 1929 (Stone 1997: 100). The production schedule of *Amazing* was slower than Gernsback's new magazines, and Gernsback had Stone's work at hand when he needed to go to press. 'Men with Wings' was Stone's first story published with Lasser, but it reads like a traditional story of super science with only faint hints of any feminist sensibility. 'Men with Wings' posited a future that was not exactly ideal for women. The story begins with sensational news headlines that scream in all capital letters, 'NORDIC FEMALES UNSAFE IN LATIN AMERICA! MANY OF AMERICA'S FAIREST HAVE VANISHED WITHOUT A TRACE!' (Stone 1929: 59). The women are being captured by a scientist named Howard Mentor, the leader of a group of scientists living in the jungles of Brazil and Peru. Mentor uses these 'Nordic Females' in his successful experiments attempting to produce winged people. The method used is vague, but the scientists in the Mentor group included 'some evolutionists', and Mentor perfects a method of injections and grafts that induce women to give birth to winged children

(68). Men taking women by force was a key aspect of Darwin's vision of sexual selection that Darwinian feminists subjected to critique, but in this story Stone does not provide much of a critical perspective on the practice. Stone also reinforces contemporary beliefs about white racial supremacy; as one character notes, the Mentorites began stealing beautiful women and scientific men to keep from too much inbreeding, but 'they refuse[d] to breed with any but people of their own race' (70). By the end of the story, these winged white people conquer the world and universal peace ensues. Stone's use of eugenics, racial hygiene and evolution to 'improve' humanity in this story is largely uncritical.

Stone extended this world in her 1930 story 'Women with Wings' which was published in the May issue of *Air Wonder Stories*. However, in this story Stone began to provide a critique of the masculinist world of the Mentor in a manner that was consistent with Darwinian feminism. Picking up centuries after the winged 'Mentorites' have conquered the Earth, these post-human beings are faced with a disease that threatens their survival. The new Mentor tells the assembled world leaders at the beginning of the story that,

> our medical men have been seeking to fight the menace that is threatening to depopulate our world of its women. Ever since the inception of the Winged Race, in fact, Science has been trying to save the women who so valiantly carry on the tremendous task of bringing young into the world. For, as you know, at least five in every ten die in giving birth to our winged children. Thus man has been able to improve upon Nature in one way, only to fall down in another. (Stone 1930: 985)

Stone clearly articulates here that the supposed progress of the eugenics-driven Mentor in her earlier story has in fact proven to be another example of scientific masculinity overreaching in a way that has dire consequences for women. Characterising childbirth as 'valiant' and framing it as a lynchpin in the species' evolutionary survival was consistent with the speculative visions of Stone's Darwinian feminist predecessors such as Charlotte Perkins Gilman and Inez Haynes Gillmore, as well as her contemporaries such as M. F. Rupert and Lilith Lorraine. Indeed, the entire story revolves around how to solve this evolutionary crisis, underscoring the importance of a women's issue as key to scientific progress.

As she develops this feminist theme, Stone is somewhat critical of the white supremacy of her earlier story. In the future of 'Women with Wings', global leadership includes 'representatives of the various races of the Earth', proving that, 'Color was of no consequence to these broad-visioned men' (986). However, in spite of trying to create a sort of rainbow utopia where eugenics has perfected every race – and where all races live in harmony – Stone still lapses into problematic stereotypes. One solution to the problem facing Mentorite society is proposed by President Chang of the Chinese Empire, who promotes a return to the raiding of the original Mentor's colony: 'There has been talk, Mentor, of sending out an expedition to other planets in search of more women! ... This is not the time for ethics, Mentor. The first of the *alated* did not balk at stealing women so that they could continue to propagate the race!' (986). Putting this morally repulsive practice – the apotheosis of predatory Darwinian masculinity – in the hands of a Chinese man associates his race with the 'savage' practices of the past. In her descriptions of the races, Stone also assumes a suspiciously Anglo-centric ideal of beauty that eugenics has helped each race to achieve: 'Gone was the racial flatness of the Chinese face and its bland inscrutable eyes, while the Negro had as regular, clear-cut features as the Englishman who sat beside him' (986). Associating 'flat' faces and 'inscrutable eyes' with the Chinese simply reinforced the essentialist racial stereotypes of the period. The supposed 'racial' facial features of both the Chinese and the 'Negro' races are explicitly represented as unattractive and undesirable by Stone, whereas the 'clear-cut' Anglo facial features are represented as an ideal of beauty other races strive to achieve. Earlier in the story, Africans are described as 'a race of people who have heretofore proven themselves extremely hardy, with only three deaths out of ten women who go through childbirth!' (985). This belief that African women are 'hardy' and white women are frail is consistent with the racial dynamic of maternity that was common in southern literature at the time. These essentialist stereotypes severely limited Stone's attempt at imagining a racially progressive future.

The strength of Stone's critique in 'Women with Wings' comes from her depiction of a strong Amazonian species. The Mentorites, instead of having to resort to stealing wives, are saved by this matriarchal species when they arrive from Venus. At first, they appear to be invaders from space who land in a mysterious cylindrical ship. The unseen inhabitants of the ship manage to stun and steal thousands of Mentorites and leave for their home planet. The Mentorites quickly build a fleet of ships, pull weapons out of their museums and launch a rescue mission. Upon reaching Venus, a mysterious voice on the radio tells them their antagonists are called 'Zoldans' and gives them instructions on how to attack. The leader of the Mentorites give orders 'to stun, not kill' when he sees 'that the Zoldan warriors were women!' (995). What ensues is a battle of the sexes which leaves a laughing 'Professor D'Arcy' doubting

> that a stranger battle has ever been enacted in the history of creation! Men and Women in battle, each trying to beat the other, yet each trying to save the other for its own benefit! Each wanting victory, but not wishing to sacrifice the other to gain its point! Without a doubt we need each other ... (995)

This battle of the sexes, in other words, is over who gets to control the heterosexual economy of sexual selection. Though the men of the Mentorites are winning the battle, the Zoldans do not surrender. Instead, their queen Waltia calls a truce, allows the Mentorites to land and negotiates a peace with her antagonists. In this way, Stone populated her most feminist story to date with powerful, Amazon-like angels who show post-humanity how natural reproduction with women as equals is better than Frankenstein-like attempts to control life. She preserves the concept that males are better evolved for fighting, but posits that it is females who are the wisest and who have the best solutions for how to progress.

Stone's representation of Zoldan society has many of the utopian elements of Darwinian feminism. The Zoldan women are represented as superior to their men, and this inversion of Darwin's sexual hierarchy is interesting in that it calls attention to the contingent nature of evolutionary development: on Venus, 'from the very beginning of time the female has always been the dominant sex, the male being accepted by them only as a biological necessity. As an intelligent being he is practically nil' (999). This matriarchal evolution echoes Gilman's Herlanders, who have evolved past needing males, and closely resembles the 'breeding males' status of men in the work of Stone's contemporary Rupert. The Zoldans have evolved from flying fish, and the women are described as 'a little less than six feet in height' and covered in scales, with wings that 'were not half so long as Earthly wings', with 'slender and deep-chested' bodies and 'the broad shoulders of a swimmer' (996). Their powerful bodies also are 'without breasts' (996), and instead of giving birth, the Zoldan women lay eggs that 'have to be incubated in the water' (999). This pattern of reproduction bears a striking resemblance to the Martians of Burroughs's John Carter novels, where the females lay eggs in sealed chambers instead of giving live birth. Stone adapts Burroughs's evolutionary model to address a particular concern of women, as the reproductive biology of the Zoldans provides a solution to the Mentorite problem of women dying in childbirth (just as the Mentorite men provide the answer to a disease killing off the weaker Zoldan males). Once the species interbreed, the combination of Zoldan mothers and Earth fathers allows all babies to be hatched from eggs. Thus, Stone's story provides a unique solution to male control of sexual selection and the dangers of childbirth, two central concerns for Darwinian feminists. While staying safely within a heterosexual economy, the conclusion of Stone's future gives women as much power as men. After a few generations of successful breeding, the story ends with the newly fused winged people of Earth and Venus considering the conquest of Mars. As in Gillmore's *Angel Island* and Irving's 'The Moon Woman', the new social situation created by enlightened men embracing flying feminist women provides benefits for both men and women and leads to a progressive and interracial future. Again, it is female choice in the selection of mates and female power over sexual selection that is key to all of this progress. The ending also implies that colonialism is not inherently wrong as long as there is balance between the genders, something that is at odds with many elements of the story and contradicts most of her other work.

In a later short story 'The Conquest of Gola', which Stone published with Lasser and Gernsback in the April 1931 *Wonder Stories*, she pushed her Darwinian feminist version of SF to its apex. The story is told from the perspective of 'the Matriarch' of Gola, which is a 'cloud-enclosed' planet in a system of 'nine planets' (Stone 2006: 36). This thinly veiled reference to Venus once again shows how Stone's stories provide a feminine version of Burroughs's masculine Mars-centred adventures. In this case, Stone tells her entire story from the perspective of a female narrator who is the only person old enough to remember the attempted invasion by the 'man-things' from 'Detaxal', which is 'the third planet from the sun' (36). Stone thus inverts the colonial gaze: instead of giving a standard story of colonisation, where Earth men conquer some planet filled with exotic and inferior aliens,

Stone gives a story from the perspective of aliens complete with the names they use for themselves and for Earth. As a part of this inversion, Stone uses a number of satirical elements to provide a biting critique of the scientific masculine gaze: the bodies of the Detaxalans are described by the Matriarch as 'a patch work of a misguided nature' (38). She says that their bodies are 'horrifying', with 'a bony skeleton somewhat like the foundations upon which we build our edifices' and exposed 'pinkish-brown skin' that has 'neither fur nor feathers' (39). The Matriarch gives these descriptions of human bodies as 'proof of the lowliness of their origin' (39). Through this inversion, Stone provides a classic example of cognitive estrangement, where the fantastic setting and perspective of the story allows for a critical distance from the discourse of nineteenth-century racial science and anthropology. Applying this discourse to the would-be colonisers from Earth calls attention to the Eurocentric conceits of both traditional scientific masculinity and SF narratives of exploration and conquest. Stone extends this critique by describing the bizarre alien bodies of the Golans using the language of superiority and beauty that was a regular feature of Eurocentric scientific discourse. The Matriarch contrasts the ugly Detaxalan bodies with 'our fine circular bodies, rounded at the top, our short beautiful lower limbs with the circular foot pads' and 'our beautiful golden coats, our movable eyes' (40). These women are not the exotic naked beauties of Burroughs's Mars, but truly alien females who find human men as repulsive as the men find them.

Stone uses distaste for conquest as an important marker of the female Golans' superiority over the male Detaxalans. The narrator relates that,

> the ignoble male creatures, breed for physical prowess, leaving the development of their sciences, their philosophies, and the contemplation of the abstract to a chosen few. The greater part of the race fares forth to conquer, to lay waste, to struggle and fight as the animals do over a morsel of worthless territory. (37)

Here Stone draws attention to the distinction between scientists and manly action heroes, with the violence and acquisitiveness of the latter seen as unproductive and animalistic. By contrast, the female Golans embrace peace, sensible development and a sense of place:

Long ago we, too, might have gone on exploring expeditions to other worlds, other universes, but for what? Are we not happy here? We who have attained the greatest of civilizations within the confines of our own silvery world. Powerfully strong with our mighty force rays, we could subjugate all the universe, but why? (37)

Stone shows the Golans as both intelligent and wise, with their technologies used to improve life and happiness in their home instead of spreading cruelty and subjugation abroad like the unintelligent men from Detaxal. In effect, the Golans have expanded their domestic sphere to the entire planet, but have no desire to move beyond their wonderful home. The mental superiority of the peaceful Golans is further contrasted to colonising stupidity during the account of the first Detaxalan visit. The men are presented to Queen Geble, who attempts to communicate with them telepathically but finds that they have 'savage' minds that contain 'a very low grade of intelligence' (41). Geble dismisses 'them from her mind as creatures not worthy of her thought', but the Detaxalans persist, asserting that they come 'with the express purpose of exploration and exploitation' (40, 42). Having already conquered Mars (which the narrator calls Damin), they make plain that they will use force to make 'Gola become one of us' in 'the Federation' that they are creating (44). When the Detaxalans launch an attack from the air with their flyers, the Golans use beams that bring the flyers to the ground and hold 'every living thing powerless against movement' (45). The Golans, of course, are immune to the effects of the beam because of 'the strength of our own minds' (45). The Golans take a few Detaxalans out of their ship for scientific experimentation or to make into slaves, obliterating the rest. Through the ease of the Golan victory, Stone shows the superiority of their anti-colonial scientific femininity.

This battle of the sexes is not over, however, and unlike 'Women with Wings', the Golans do not attempt a truce with the Detaxalans. Through duplicity and intrigue, the second wave of Detaxalan ships is temporarily able to subdue the Golan women. This is made possible by the weak, stupid males of Gola who have been seduced by tales of power from the remaining Detaxalan male slaves. The narrator says that her Detaxalan slave Jon subdued her in her sleep, and there is 'a moment where a new emotion' takes hold of her: 'the pleasure to be had in the arms of a strong man, but that emotion was short lived'

(47). As Larbalestier notes, the Golan narrator temporarily falls for 'the romance discourse' before it is 'dispelled almost immediately' (2002: 11). In this way, Stone satirically tweaks the standard Burroughs vision of beautiful, exotic women who swoon in the arms of powerful men such as John Carter. Instead, she has a bizarre alien woman quickly regain power and reject the romantic possibilities of Detaxalan men along with their 'commerce and trade, business propositions, tourists and all of their evil practices' (47). Through this language, Stone links capitalist exploitation with masculine colonial conquest, rejecting both in favour of harmonious domestication of a home planet. The name of the story is itself satirical, with the 'conquest' in fact a temporary setback for the Golans. The story ends with the men of Detaxal defeated completely in a way that challenged the sexist assumptions of many science fiction stories of the day. Stone continued to criticise colonial exploitation in her work, telling tales of interplanetary adventure with a critical edge. After 'Conquest of Gola', however, she never returned to such an explicitly feminist plot told from the perspective of a superior female narrator. She continued to publish SF for the remainder of the decade, but after Lasser left the field she found editors increasingly hostile to her work.

Lilith Lorraine

Lilith Lorraine was perhaps the most politically radical of the women who wrote fiction for early SF magazines, selling four unabashedly socialist stories to Gernsback's publications between 1929 and 1933. The first of these, 'The Brain of the Planet', was published as the fifth issue of Gernsback's *Science Fiction Series* in October 1929. *Science Fiction Series* was 'a set of booklets, each containing twenty-four pages' that were released in groups of six (Ashley 2004: 160). They were advertised in Gernsback's other SF magazines and sold well, providing yet another avenue for women such as Stone, Lorraine and Amelia Reynolds Long to get their SF into print. Lilith Lorraine was the pen name of Mary Maude Wright, a teacher who was educated in Arizona and California and held a number of other jobs – and pursued a number of other SF ventures – during the course of her life (Yaszek and Sharp 2016: 106). 'The Brain of the Planet' established a number of Darwinian themes and feminist ideas that would be consistent throughout Lorraine's career. The first of these is telepathy, with Lorraine following H. G. Wells and Clare Winger Harris in representing this mental power as a likely product of humanity's current evolutionary path. In 'Brain', the chairman of the 'Arizona Institute of Applied Psychology' announces the discovery that 'telepathy is a *fact*', but 'the twin forces of heredity and environment' have led human minds to be 'fiercely protective of ... individuality' (Lorraine 1929: 4). Using the analogy of radio, he explains that human brains are not 'in tune' with one another (4). The chairman also gives voice to the common 1920s fear of radio-like technologies bringing minds into tune, arguing that leaders 'have already wrought irreparable havoc among men' without access to such mind-altering gadgets (5). The young secretary of the institute Harry Maxwell argues that using telepathy to control minds would cause harm 'only because economic conditions under the competitive system are such that it *pays* far more ... to exploit our fellows than it does to use that will for the common good' (5). Here Lorraine makes plain her Darwinian feminist version of socialism, where the supposedly feminine trait of cooperation is seen as beneficial and the masculine trait of competition is associated with capitalism, selfishness and harm. The possibilities of mind-control technology allow her to construct a story where socialist revolution is non-violent, and where the world transforms into a utopia. Lorraine echoes the utopias of Lane and Gilman by making feminine cooperation central to her society, though Lorraine's feminism became even more explicit and pronounced in her later publications.

Like her feminist predecessors, Lorraine filters her story through the perspective of a male character, the young psychology professor Maxwell (though he is not the narrator). Lorraine's utopia takes shape when Maxwell travels 'into the wilds of Mexico' to a remote mountain with a mechanic named Jerry Brand (7). What he constructs there is 'the brain of the planet', a radio transmitter enabled by 'the evolution of the brain' that provides 'a centralized intelligence to direct the rudimentary ganglionic centers that pass for the brains of human beings' (8). As Maxwell explains to Brand, 'through lack of a super-brain, the disconnected and crystallized brains of individuals are perpetrating fossilized and primitive institutions long after these institutions have outlived their usefulness' (9). Instead of the status quo – where humanity is enslaved by capitalism and forced into competition – Lorraine has her scientist enslave humanity into cooperation and create a kind of 'feminist utopia' (Donawerth 1990: 254). In the hands of most SF authors, especially women who are critical of scientific masculinity, this would provide the formula for a villainous mad scientist who is over-reaching by trying to strip humanity of its free will. This is precisely what Harris warned against in her story 'The Evolutionary Monstrosity'. Lorraine, however, rarely represents science and scientists in this way. Instead, she places this masculine scientific power in the hands of a benevolent socialist who brings about a socialist revolution against the capitalist 'Masters of the Machines' that eventually benefits women as much as men (Lorraine 1929: 10).

In fact, Lorraine goes so far as to represent this as returning humanity to its natural course of evolution. Maxwell notes that the Masters of the Machines 'work *against* human progress', and his plan will take a 'sea of mental stagnation' and 'turn its currents in the *direction* of evolution and open pathways of new ideas into the limitless ocean of thought' (11, 12). The slippage here between evolution and progress was common in early SF, but Lorraine clearly points out the contingent nature of evolution and what she counts as progress. Once the machine is working, this vision of progress becomes more fully developed in Lorraine's utopia. Like most utopias, however, this one is built on exclusion and death in spite of the purported non-violent nature of the revolution. When the machine is turned on and beams 'rosy rays' of 'love', the minds of people with 'known radical tendencies' and those 'hard conservatives ... known for their unswerving devotion to the creed of Mammon' become consumed by psychological diseases and die (15). In short order, 'the mighty energies of man's superhuman machinery were enlisted on the side of human progress' and the workday is reduced for all peoples with each new invention (16). A 'World State' arises very quickly, and with enough 'leisure' humanity begins 'to bring its mental evolution up to the level of its mechanical evolution' (18–19). When Maxwell finally turns off his machine, the freed 'primeval impulses' lead to a 'last stand of the old savagery' that ultimately cannot overthrow a 'civilization' where people have 'become accustomed to leisure, to freedom, to equality of opportunity' (20, 23). Lorraine repeats the Victorian formulation of 'savagery' and 'civilisation', reinterpreting this hierarchy of cultures (and implicitly, races) as a struggle between capitalism and socialism. Lorraine's feminist sensibility remains relatively coded in the story, with vague references to 'equality of opportunity', but her later work would make her feminism much more explicit.

David Lasser clearly approved of the explicit socialism in 'Brain' and published her more elaborate feminist utopia 'Into the 28th Century' in the November 1930 issue of *Science Wonder Quarterly*. This story constituted Lorraine's most traditional and fully formed utopia during this period. Following the model of earlier Darwinian feminists, a sympathetic man narrates the story. In this case, he is a recently discharged navy man named Anthony who is pulled into a future where feminism has helped create a perfect world. Anthony describes the people who pull him to the future as,

> young men and women with hair every color of the rainbow – red, green, yellow, purple, and intervening shades unknown to me. Both sexes had curls falling to their shoulders, but the boys, notwithstanding, did not convey the impression of effeminacy. They were perfectly formed, athletic, muscled and gracefully lithe. The girls were a combination of Venus de Milo and Diana the Huntress. (Lorraine 2016: 111)

As Gilman did with her Herlanders, Lorraine stresses the physical vitality of the female denizens of 'Nirvania'. However, Gilman reinterpreted the legend of the Amazons in a way that presented them as non-violent and – to the eyes of the male narrator – lacking in femininity. This was Gilman's vision of women freed from the constraints of sexual selection, and how women would evolve once they no longer had to please men to survive. Lorraine presents a society where gender differences are enhanced rather than minimised. Anthony's love interest Iris explains that,

> Man still leads in invention, mechanics, mathematics, and the more strenuous sports. Woman has ceased to imitate man, being content in her own sphere. She has intensified her femininity, wherever it can be done without a sacrifice of her health and freedom. Thus she has preserved that pleasing contrast in the sexes which perpetuates their appeal for one another. (122)

Where Gilman's Herlanders were proficient with science and technology, Lorraine uncritically repeats Darwin's formulation of man the toolmaker and rejects women's pursuit of such masculine domains. Lorraine also celebrates the heteronormative romance of Darwinian evolution, presenting it as natural and one of the most sacred aspects of life that must be preserved as humanity evolves. Viewing the expansion of gender differences as a positive development was an idiosyncratic element of Lorraine's story, especially in relation to other Darwinian feminists and women SF writers of her day.

However, Lorraine shared a critical view of patriarchal privilege that was very similar to other Darwinian feminists. In describing the revolutions that shaped her utopia, Iris provides a Darwinian feminist critique of gender relations in the twentieth century:

> The knighthood of the ages was but a sugar-coated pill that concealed a soul-killing poison, an opiate to drug the intellect. This discovery embittered woman and almost caused a war among the sexes. The ultra-feminists began to acquaint the world with the true status of the case. They began to demand chivalry, but not as an opiate to lull the reason into submission to a sex whose last claim to superiority had been undermined. They demanded it as a tribute still due to those who were *more* than equals, because of the sacrifice they still endured in giving birth to man. Man met this demand with taunts and insults. Woman gave him his choice between the restoration of chivalry and the surrender of his ancient privileges, even the surrender of his parenthood. She grimly stated that it were better for humanity to die painlessly through the ceasing of birth than to commit suicide through the continuance of man-made institutions. (121)

Following other Darwinian feminists, Lorraine connects feminist progress with women gaining control of sexual selection. In the eugenics-driven logic of Lorraine's story, women controlling reproduction provides a key for improving the race both biologically and spiritually. Eliza Burt Gamble's *The Evolution of Woman*, Inez Haynes Gillmore's *Angel Island* and Gilman's *Herland* all shared this view that motherhood was a central pillar of civilised progress, and that biological improvement hinged upon women being freed from male ownership of their bodies. As in Gillmore's *Angel Island*, Lorraine's '28th Century' has women withhold sex in their fight to gain the power of choice. Women do not literally fight as they do in Stone's 'Women with Wings', but rather play on the love, desires and sympathies of men in order to be emancipated.

Another significant feature of Lorraine's utopia in '28th Century' is how she reimagines a number of common problems facing women of the twentieth century. Chief among these is the process of pregnancy and giving birth. The love-smitten narrator has this explained to him by the 'green-haired sea goddess' Iris (117):

> Birth is entirely different from the horror that it was in your day. The embryo is removed from the womb shortly after conception and brought to perfect maturity in an incubator. The old relations of the sexes except for purposes of procreation have practically ceased. The great energy back of the wasteful reproduction that led at last to death, has been turned into the channels of rejuvenation. (124)

Like Stone's 'Women with Wings', Lorraine imagines here that the pain and dangers of childbirth are prioritised by society and solved scientifically. This feminist form of progress – where women's issues become the focus of scientific advancement – was a common feature of SF written by women in this period. Lorraine's language regarding sex as a waste of energy was another hallmark of Darwinian feminism, and closely mirrored language in the writings of Gamble and Gilman. It looks to the model of nature Darwin described in his chapters on sexual selection in non-human animals, and imagines a return to 'natural' cycles of reproduction where sex is only for purposes of procreation and the sexual desires of men are seen as unnaturally exaggerated. As in 'Brain', Lorraine has thoughts and telepathy serve as a means for social harmony, in this case between lovers: instead of physical sex, lovers connect telepathically and 'the divine ecstasy generated by our love for one another is transmuted to the higher planes of soul expression' (124). Rather than using telepathy to enslave others, Lorraine imagines that it allows for the most powerful form of intimacy in the next step of the brain's evolution by transmuting or elevating physical orgasms into an ecstasy of the soul. Lorraine's utopia also includes people of both sexes taking 'turns at performing essential household tasks which by reason of the marvelous labor-saving devices, are rendered extremely light' (130). As in Minna Irving's 'Moon Woman', technology and equality has freed women from domestic servitude, and they can fly through the air 'equipped with artificial wings of gorgeous plumage' (128). Like the flying women of Gillmore and Stone, Lorraine has her flying women serve as icons of emancipation and Darwinian feminist progress.

Lorraine's vision of eugenics as necessary for progress takes an interesting form in her representation of race. As Iris explains,

> we had weeded out undesirable racial strains by wholesale sterilization. The carefully preserved superior strains in the various races have united to form a super-race. Children are still born occasionally whenever a marriage is consummated wherein the contracting parties possess qualities of genius that we desire to see multiplied. (124)

Again, Lorraine's story engages in a type of evolutionary essentialism, where each sex and race has an essence that can be modified into a more utopian form. Lorraine posits that the best of each sex can be enhanced through eugenics, and the best of each race melded to form a kind of rainbow eugenics that attempts to reject the idea of white supremacy. When describing the mayor of Nirvania, the essentialist nature of Lorraine's racial imagination is obvious:

> He was tall, lithe, and splendidly proportioned, as were all this super-race. His features portrayed a subtle blending of the noblest qualities of the ancient Greek, the North American Indian, the Oriental, and the Anglo-Saxon. The aquiline nose was unquestionably Grecian, the high cheekbones were those of the Indian, the soft, mysterious eyes held all the charm of Hindustan ... (129)

The orientalist language of this passage undermines Lorraine's attempt to imagine a future for race that is different from Gilman's Aryan paradise. By associating certain facial features with specific races, Lorraine engages in a type of essentialism that runs counter to her attempt to tweak the racial egotism that she saw as a blight on society. There is also no mention of Africans in this future super-race, implying by their absence that there was nothing in their race worth preserving.

The SF stories of women such as Harris, Stone and Lorraine demonstrate many of the hallmarks of Darwinian feminism, including their essentialist formulations of gender, sexuality and race. Seen through this lens, their stories provide understudied examples of how women were wrestling with both the scientific debates and political movements of their day, and extending feminist arguments of the 1900s and 1910s. This is one of the unique features of women's SF in the early specialist magazine era, and magazines such as *Amazing Stories* and the various iterations of Gernsback's *Wonder Stories* served as a welcoming platform for such politically engaged feminist visions of the future. When Lasser left in 1933, however, SF magazines became decreasingly amenable to Darwinian feminist speculation, and women SF writers had to adjust their work to fit into a rapidly changing editorial landscape. Some such as Stone faced outright sexist rejection in spite of their extensive résumés, a fact that led them to leave the field altogether. Others adapted their style to mirror their male counterparts more closely, and to appease the more conservative sexist attitudes of a new generation of editors.

5 Darwinian Feminism and the Changing Field of Women's Science Fiction

As a major scientific idea, evolution either explicitly or implicitly constituted the framework for most extrapolation within SF. Narratives of techno-scientific progress took their shape and language from Darwinist discourse, with toolmaking humans developing civilised, galaxy-spanning enterprises or devolving into brutish savagery according to a colonial paradigm Darwin established. Darwinian feminism was just one particular form that such extrapolations took, a form that was consistent with the editorial policies of Hugo Gernsback. A long-standing advocate for scientific literacy, Gernsback saw such 'scientific fact' as central to the vision he spelled out in his first editorial on SF in the April 1926 debut of *Amazing Stories* (Gernsback 1926: 3). According to Gernsback's vision, stories should contain 'lengthy and detailed explanations of current scientific knowledge and discoveries' along with a fantastic and adventurous plot (Westfahl 1998: 39). This included fantastic inventions, with Thomas Edison as the popular model for the 'young US male inventor hero' whose gadgets and ingenuity help him in his adventures (James 1994: 26–7). In his earlier magazines such as *Modern Electrics* (1908–13) and *The Electrical Experimenter* (1913–29), Gernsback wrote and published his own stories of heroic inventors. Gernsback published these magazines in part to promote his electronic parts importing business by educating and inspiring amateur radio enthusiasts (and, later, television enthusiasts) and showing them what science and technology might bring them in the future. As Gernsback's vision was debated and revised over the next fifteen years, women made a number of significant contributions to the use of science in magazine SF. Many embraced Darwinian feminist plots and issues as a part of this scientific extrapolation, but as the 1930s wore on and magazines changed editors, such feminist perspectives became less welcome.

The changing editorial landscape of the 1930s

Early in Gernsback's run at *Amazing Stories* it became apparent that many readers valued scientific accuracy at least as much as the editor. Gernsback encouraged fan debate about scientific plausibility by publishing letters in each issue showing how his readers wanted 'the scientific content' to be 'both valid and essential to the plot' (Attebery 2002: 40). Gernsback took the additional step of hiring Dr T. O'Conor Sloane, a retired professor of natural science at Seton Hall, to serve as managing editor, and in later magazines included a list of 'nationally-known educators' in a dozen areas of natural science and engineering who worked with the magazine (Gernsback 1930: 5). When Gernsback lost control of *Amazing Stories* in 1929, Sloane took over operations of the magazine. In his first post-Gernsback editorial, Sloane articulated his own version of Gernsback's vision, saying that the 'basic idea' of the magazine was to publish 'fiction, founded on, or embodying always some touch of natural science' (quoted in Westfahl 1998: 165). However, Sloane was more interested in established science than in 'prophetic vision', and seemed dubious about Gernsback's belief in the value of SF to inspire scientists and inventors (Westfahl 2007: 22). Sloane was particularly sceptical of the feasibility of space travel, and his editorials tended to read like Victorian-era lectures by a professor of science (which indeed he had been). Sloane noted in a November 1929 editorial that 'since our readers like interplanetary stories ... and since there is any amount of science, mechanical, astronomical and other to be gleaned therefrom, we certainly shall be glad to continue to give them, even in face of the fact that we are inclined to think that interplanetary travel may never be attained' (Sloane 1929: 677). Though he did publish space flight stories, he did so for entertainment purposes and to teach science, and not because he thought it would inspire scientists to do something he thought was impossible. Sloane and his managing editor Miriam Bourne continued to publish stories by women such as L. Taylor Hansen, Leslie F. Stone, Minna Irving, Clare Winger Harris and Amelia Reynolds Long.

Gernsback's emphasis on the science in SF stories also influenced multi-genre magazines such as *Weird Tales*, which had been publishing SF since its first issue in 1923. Under editor Farnsworth Wright, *Weird Tales* had published 'lost race' stories, 'monster-mutation' stories and 'interplanetary' stories before Gernsback even founded *Amazing*

Stories (Ashley 2000: 43). When *Amazing Stories* hit the news-stands, however, some fans of *Weird Tales* began to complain about including SF in the magazine, instead encouraging Wright to stick to genres such as fantasy and supernatural mystery. In developing a clearer definition of SF for *Weird Tales*, Wright asserted that the magazine would continue to print 'the cream of the weird-scientific fiction that is written today', which he characterised as 'tales of the spaces between the worlds, surgical stories, and stories that scan the future with the eye of prophecy' (quoted in Weinberg 1999: 120). Wright quickly turned this into an advertising blurb for subscriptions to the magazine that ran in *Weird Tales* and some of its sister publications from 1928 through the early 1940s. Where Gernsback's magazine had the masthead slogan '*Extravagant Fiction Today – Cold Fact Tomorrow*', Wright declared to his readers, 'The pseudo-science of today is the real science of tomorrow' (quoted in Weinberg 1999: 120). Wright continued to publish 'weird-scientific' stories throughout his tenure, and seemed particularly receptive to stories by women that included a strong element of horror with the SF.

Astounding Stories, the strongest direct competitor of *Amazing* that emerged during the early 1930s, hewed closer to Gernsback's vision of SF than Wright in some ways. However, *Astounding* was similar to *Weird Tales* in that it put more emphasis on adventure, devoted less space to scientific explanation and embraced the occasional story based on pseudo-science. Under editor F. Orlin Tremaine, who edited the magazine from October 1933 to October 1937, *Astounding* promoted the new 'thought variant' SF story 'whose driving force was a speculative idea, rather than a gadget or a sequence of cliff-hangers' (James 1994: 49). Where Gernsback and Sloane emphasised natural science and engineering, Tremaine's thought variant story allowed for authors to draw more freely from social sciences such as psychology, sociology and anthropology in developing complex characters and detailed alien cultures.

Throughout the late 1920s and 1930s women authors were in the middle of the innovations and transformations that expanded Gernsback's original vision for science in SF. Clare Winger Harris was the first to prove that women could contribute to the genre in SF magazines even with Gernsback's emphasis on scientific facts and technical plausibility. Harris and L. Tayler Hansen were popular authors who exemplified the best of the Gernsbackian tradition; however, their work also occasionally employed strategies consistent with the feminist SF tradition of Mary Shelley and Charlotte Perkins Gilman, using male narrators to critique masculine science while preserving a sense that progress can benefit humanity. Darwinian feminism flourished most when Lasser was Gernsback's right hand from 1929 to 1933, but the authors Lasser promoted such as Leslie F. Stone and Lilith Lorraine continued to publish on similar themes and in similar styles after he left the field. Original *Astounding* editor Harry Bates did not publish much work by women, though stories by Sophie Wenzel Ellis and Lilith Lorraine found their way into his magazine during his three years at the helm. C. L. Moore published SF in *Weird Tales* that mixed horror with SF in a manner that brought different perspectives on gender, sexuality and science to the fore. Tremaine published fiction by Moore, Kaye Raymond and Amelia Reynolds Long. By the end of the 1930s, however, women such as Stone and Lorraine left SF in frustration in large part because of the changes new editors brought to the specialist magazines. Moore created new kinds of Amazons and liberated monsters that were not limited by the essentialist heterosexual economy of Darwinian feminism, and that fit with new editorial visions of what SF could be. Women writers such as Moore and Leigh Brackett also began writing stories that had no feminist impulse or feminine perspective, and that simply engaged in the same discourses and story types as their male contemporaries. Women were still flourishing in magazine SF, but by the end of the 1930s Darwinian feminism was no longer a part of women's SF.

Lilith Lorraine's later work

Lilith Lorraine's most utopian work was for David Lasser, but she also placed three stories outside the publications for which Lasser worked. Lorraine published the short story 'The Jovian Jest' with upstart *Astounding Stories* in May 1930, moving away from her utopian format and appealing to a different editorial sensibility. As the name implies, the story is satirical and uses a strange being in a farmer's field to call attention to the limited reach of human knowledge. The being is a pulsating, growing, giant amoeba that shoots out tentacles to grab people who get too close. After pulling a farmer and an egocentric scientist named Professor Ralston into its core, the amoeba drains their brains of content. The amoeba then uses the professor's body to communicate with the assembled crowd, noting the professor's 'meager thought-content' is barely adequate (Lorraine 1930: 230). The amoeba describes the professor's brain as 'a lumber-room in which he has hoarded a conglomeration of clever and appropriate word-forms with which to disguise the paucity of his ideas, with which to express nothing!' (231). The amoeba proceeds to give an extended lecture about the superiority of his species and the limited nature of humanity, while encouraging humanity to continue to reach for the stars. When he returns their minds, the 'jest' of the title is that the amoeba gives the farmer the education of the professor and leaves the professor with 'the meager educational appliances' of the farmer (233). While continuing to embrace Victorian formulations of colonial expansion and progress, Lorraine in this story provides her most pointed criticism of science in its current form. The arrogance and vapidity of Professor Ralston is a stark contrast to the heroic scientist Maxwell in 'Brain of the Planet'. Her satirical representation of scientific masculinity recalls the descriptions of scientific societies in Cavendish's *Blazing World*, and bears many resemblances to the comic writing of her contemporary Amelia Reynolds Long. Like Cavendish and Long, Lorraine shows how the simple performance of masculine scientific authority was often used to overawe the uneducated and delude the educated into a false sense of assuredness and importance. She also shows the limitations of such knowledge in the face of nature and the wide universe of possibilities.

Lorraine's 'The Celestial Visitor' uses a satirical tone similar to 'Jovian Jest'. Billed as an 'absorbingly different satire' when it appeared in the March 1935 issue of Gernsback's *Wonder Stories*, 'Celestial Visitor' also provided an anachronistic representation of evolution and telepathy in a vein similar to her earlier stories (Lorraine 1935: 1190). By 1935, the managing editor of *Wonder Stories* was young Charles D. Hornig, a high schooler whose only real interest for a story was that it creates a 'Sense of Wonder' (Davin 1999: 67). 'Celestial Visitor' provided just that, focusing on the experiences of an alien named Zanor as he visits a strange planet called 'Erath'. Zanor is a young man who lives in Eutopia, a society on an asteroid near the Earth. Zanor is bored and suffers from 'too much contentment', so he begins an investigation that proves Eutopia was built by scientists from Atlantis who constructed a space ship and escaped the cataclysm they rightly predicted would destroy their homeland (1935: 1191). The Eutopians have evolved advanced forms of communication such as the 'radio-television' and the 'thought transmitter', which allow for interplanetary conversations with the people of Venus telepathically (1193). When Zanor discovers the origins of his people he decides to visit 'Erath', which was thought to be uninhabited because, 'Time and again we had laid thought-lines to Erath, our nearest neighbor in space, but they failed to respond to our vibrations' (1194). Zanor's visit leads to a classic inversion of the colonial gaze, where Lorraine uses the superior perspective of the perfect Zanor to criticise contemporary society, often with comic effect. For example, upon his arrival, Zanor walks down the main street of the small town, 'expecting to reach the home of the Head-Savage (the state of architecture having convinced me that the Erathians were in a semi-savage state, though I could little reconcile this with the high culture described by [our scholar], unless the Great Catastrophe had hurled mankind back into barbarism)' (1196). This passage also lampoons the language of colonial anthropology and evolution by putting the Americans at the bottom of the hierarchy when compared to superior, evolved, super-civilised aliens.

Like Clare Winger Harris, Lorraine follows a common theme from domestic fiction: Zanor is the perfect man who appreciates a strong intelligent woman. After his arrest, Zanor falls in love with a young feminist who he calls 'The Advance Type'. When he is later seated next to her at dinner, he is 'charmed more and more every moment by her beauty and intelligence' (1203). By using a male narrator, Lorraine imagines the perfect sympathetic man who sees activist feminists as more evolved and as the most desirable. Zanor and his people also provide an alternative to the horrors of sexual selection and women's oppression in marriage: Zanor narrates that,

> Our elders tell us that when a man has found his eternal mate a great spark leaps between them and that they then know that the Indissoluble Bond has been formed, the bond that knows no breaking and that no Eutopian ever dreams of breaking, since, being led to our mates by love alone, and by no outward compulsion, there can never be a mistake. Never yet had I felt in the presence of a woman as I felt today, today when I knew too well that the Great Spark had leaped. (1203)

For Lorraine, Zanor is the perfect man because he has absolute fidelity, and comes from a society that has eliminated the last vestiges of oppressive sexual selection diagrammed by Eliza Burt Gamble and Charlotte Perkins Gilman. More importantly, his telepathy is something that leads to perfect honesty. When asked, 'Have you never on your planet heard of the phenomenon called a lie – a false statement?' Zanor replies, 'Of course not. What object would there be in perverting the truth since all minds are open to one another?' (1200). The prospect of marriage to Zanor is made more attractive for 'The Advance Type' feminist by the fact that the Eutopians have transformed childbirth: they follow the 'custom of allowing only a few children to be conceived each year and brought to maturity in incubators' (1203–4). As in her earlier 'Into the 28th Century', Lorraine echoes the speculative visions of Darwinian feminists such as Gilman and uses SF to imagine freedom from the pain and serious health risks posed by childbirth. In effect, her version of 'Eutopia' again centres on overcoming the 'brutal' nature of men and eliminating the dangers of marriage and reproduction. In this way, Lorraine was clearly in synch with her contemporaries such as birth control pioneer Margaret Sanger, another Darwinian feminist whose work Gernsback promoted in other venues.

Like 'Celestial Visitor', Lorraine's 'The Isle of Madness' was published by Gernsback's *Wonder Stories* in 1935 under managing editor Hornig. Lorraine's familiar themes are present, but the plot of the story is almost an inversion of 'The Brain of the Planet'. The story posits insanity spreading through a world where humanity's machines have 'become [their] master' and automation leads to unemployment instead of 'the hours of labor' being 'lessened' (Lorraine 1935: 653). Where 'Brain of the Planet' imagines a saviour scientist who rescues humanity, 'Isle' imagines a world that is taken over by a mind-controlling 'thought-machine' possessed by the 'Iron Masters' (653). Their machine cuts off the 'higher impulses' of the mind, leading to a devolution and a 'narrowing racial intelligence' (654). Those with evolved minds that can 'withstand the diabolical hammering of the great machine' are rounded up into 'mad-houses' that eventually become concentrated on an island (654). Narrated by Ulthor, a descendant of the mad islanders, the story chronicles how they had to go through an evolutionary process similar to 'Americanization' on the frontier described by Frederick Jackson Turner (Sharp 2007: 59–62): the islanders are 'forced to relinquish the artificial comforts of the Machine World' and emerge from the struggle to survive with an even better civilisation than the one they gave up (Lorraine 1935: 654-5).

However, as in 'Into the 28th Century', Lorraine goes out of her way to imagine how the future can be interracial. The island where the 'abnormals' are deposited had 'long been inhabited by a superior race of savages, the only primitive strain left on earth. Fortunately, their mental capacity was as great as that of the much vaunted Caucasian race; for somehow, the brain-stuff of the ages, the raw material of the mind, is no respector of races' (655). The islanders, of course, 'produced a superior human strain' by mating with these natives (655). Here Lorraine repeats the essentialist claims about 'savagery' that are based in the comparative method, an early anthropological approach where cultures were ranked along one axis of cultural and technological progress (Sharp 2007: 16–17). Lorraine also reinforces the eugenics ideal that some people are biologically superior and that the future should be shaped by superior individuals reproducing only with one another. However, she rejects the idea that one race is biological superior over another, instead asserting that the mixing of small groups from various races that have 'superior' traits can lead to a better form of (post)humanity. Lorraine also includes a feminist message in this story of utopian progress: 'Judicious birth-control prevented over-population, and woman, freed from the curse of excessive child-bearing, expanded in intellect, and hand in hand with man ... forged upward to perfect equality' (655). Lorraine thus repeats a Darwinian feminist plot point from her earlier work, where women's control of reproduction provides the key to the social and biological progress of the species.

What is unique for Lorraine in this story is her depiction of apocalypse. Ulthor leads a party to investigate the dystopian hellscape of the 'Machine Men and their civilization' only to learn that it has collapsed and all that remains of humanity is 'human wolves' who are 'strangely doubled over and with thick matted hair, writhing snake-like from awful, jutting heads' (656, 660). Ulthor and his 'Party of Expansion' have to fight off overwhelming numbers with superior technology in their attempts to bring socialist feminist civilisation to the masses devolved by capitalism to the level of 'cannibalism' (663). In one sense, Lorraine is extending her use of colonial frontier narrative, but once again she overlays this with socialist ideals: the 'civilised' socialists from the island are represented as justified in their conquest of the devolved and bestial 'savages' of capitalism. This apocalyptic warning story anticipated the many nuclear frontier stories of the post-war period, where small groups of individuals have to struggle to protect civilised ideals in the face of mutated savagery. It also provides another example of how Darwinism offered a number of narrative tools and scientific concepts that were used by women in ways that were rarely seen with male SF authors.

However, this was also the final work of fiction Lorraine ever published in a SF magazine. The spring 1943 issue of the fanzine *Acolyte* published a letter from Lorraine where she states that she left SF because 'the market' was too 'stereotyped and standardized' and limited by 'editorial fetishes' (Lorraine 2016: 315). Lasser had left the field in 1933, and Gernsback left in 1936 when he sold *Wonder Stories* to Standard Magazines. The new owners retitled the magazine *Thrilling Wonder Stories* and released their first issue in June 1936 under the editorial leadership of Leo Margulies, Harvey Burns and Mort Weisinger. These new editors aimed to try and capture a younger audience and 'bigger market' who were interested only in 'action and thrills', and jettisoned the stories Hornig had already accepted and typeset such as Leslie F. Stone's 'The Other Side' (Ashley 2004: 250–1). Stone's story never saw print, and the fall of *Wonder Stories* – along with editorial changes at *Astounding Stories* two years later – sounded the death knell of Darwinian feminist SF. Lorraine quit publishing stories in SF magazines and turned to publishing poetry, founding and editing a series of magazines and anthologies while leading the nascent 'Cosmic or Stellar poetry movement' of the 1940s and 1950s (Yaszek and Sharp 2016: 256). Stone continued to publish in SF magazines after Lorraine's departure, but found the field increasingly hostile to her work.

Leslie F. Stone's later work

After the publication of 'The Conquest of Gola' in the April 1931 issue of *Wonder Stories*, Stone turned away from explicitly feminist themes. However, her work continued to hammer away at the folly of colonial exploitation and scientific masculinity. Stone's final stories for Gernsback's *Wonder Stories* with Lasser as the managing editor were 'The Hell Planet' in the June 1932 issue and 'Gulliver, 3000 A. D.' in the May 1933 issue. 'Hell Planet' follows a crew on a mining expedition to the recently discovered 'Vulcan, innermost planet' of the solar system

(Stone 1932: 15). As Lasser's introductory blurb notes, Stone's story questions 'Man's damnable desire to conquer, to nose in where he doesn't belong' (15). This questioning of colonial conquest continued the critique of the masculine mindset Stone had begun in 'Out of the Void' and 'Conquest of Gola', but this time it had no significant female characters or any sense of a positive female alternative. The men in the story who discovered the planet Vulcan 'had died a horrible death' except for the pilot Wendell, who flew it 'back to Earth with its gruesome burden of dead men, and his tale of discovery' (15). Lured back by the promise of wealth from the rare minerals of Vulcan, Wendell recounts how his previous shipmates died after eating radium-tainted food provided by 'fox-like' natives who 'thought' the visiting humans 'were gods' (16, 18). After the second colonial mission to extract ore ends up in another death-filled catastrophe, Wendell tells the only other survivor that,

> Those poor untutored savages will fight to preserve their rights. Thousands will die before they learn their lesson; the rest will become slaves to dig out the ore ... In the future men will point to you and me. They will say ... 'those pioneers ... they were men!' Bah! Sheep! That's what we are ... pigs for the slaughter. (27)

Despite repeating the tired racist stereotype of the worshipful, simple, animal-like natives, Stone's story is consistent in presenting the colonisers as greedy interlopers who are concerned only with plundering resources. The fate Wendell envisions for the natives of Vulcan was the same fate the Arawaks and other Native Americans suffered under the '*encomienda* system' of Christopher Columbus, a 'forced-labor' system for mining resources that led to mass death for the natives while enriching the Spanish (Loewen 2007: 56–7). Stone mocks the cruelty of proving manhood by participating in colonisation, dismissing it as another illusion of a misguided masculinist culture. 'Gulliver, 3000 A. D.' was an equally grim account of a failed colonial mission blotted by masculine violence and arrogance.

Stone placed her most feminist and critical work with Lasser at the *Wonder* magazines between 1929 and 1933, but that was only one part of her prodigious output during this period: Stone sold seven stories to Sloane at *Amazing Stories* between 1929 and 1936. Most of Stone's stories were very long novellas or novels that involved sprawling plots across planets and different periods of time, and that took up a great deal of space in the issues in which they appeared. Her last sale to *Amazing* was 'The Human Pets of Mars' in the October 1936 issue. The story was typical of Stone's work that was not published with Lasser: it focused on a male hero, female characters are little more than damsels in distress and the critique of colonisation is present but muted. In other words, it was closer to the writing of most of her male counterparts. 'The Human Pets of Mars' begins with a Wells-like invasion by technologically superior 'decapod' Martians with 'flabby sock-like' bodies 'topped by a round soft head from which projected the tentacles' (Stone 1974: 730). Instead of engaging in a colonial war of extermination like Wells's Martians, Stone's gigantic decapods stroll through Washington DC like tourists and pick up a number of living souvenirs including a horse, a dog, a kitten and several people. When they return to their ship, they capture an investigating government engineer named Brett Rand, and the story shifts to his perspective as he suffers the indignities of being treated like a pet back on Mars. The story has some of the flavour of Stone's earlier criticisms of colonialism, as the human pets are treated poorly, kept on leashes and fed only 'table scraps' that make them ill (743). Stone peppers the story with explanations of astronomy and Martian geography that would have pleased *Amazing* editor Sloane. Her human captives also include a few crudely drawn black characters who are thrown in partly for comic relief, with a hysterical 'negro' woman repeatedly saying things such as, 'Hit's de punishment ob de lawd' (749). Stone thus deploys the 'clown stereotype' that was popular across media in the 1930s, a stereotype that portrayed African Americans as humorously superstitious and illiterate (Hall 1995: 22). When Isaac Asimov reprinted the story in his anthology *Before the Golden Age*, he noted that he was 'keenly embarrassed by the simple-minded portrayal of the Blacks in the tale' (1974: 773).

Asimov also claimed that reading Stone's 'Human Pets of Mars' inspired him 'for the very first time' to try and write 'science fiction' instead of fantasy (774). Asimov lauded the plot, which appealed to him because it was a 'story of indomitable men winning out over immense odds' (773). Brett and George, the indomitable men, begin to devise a plot to escape Mars and save their fellow human pets before any more die from the neglect and abuse of their cruel Martian masters.

Stone's earlier stories with Lasser mocked the pretensions of the male action hero, but in 'Human Pets' she represents Brett as taking leadership in the group 'because it seemed the only natural thing to do' (767). When he is challenged by another man, it is a female character who defends Brett's natural leadership, saying 'He's been the only man here with guts – yes, I said guts – to rescue us' (767). Brett learns how to operate a giant Martian ship and with George's aid he helps the remaining humans fly back to Earth. Along the way, they learn how to deploy the ship's shields, fire its weapons and use its navigation system, eventually winning a space battle with the pursuing decapod Martians. This plot twist caused Asimov to lament that he 'was dubious as to the possibility of Earthmen taking over completely alien vessels' (773). The heroic ending of the story – one that sees Brett marry the distressed damsel he saves – is not fully celebratory. Only a few of the humans survive the ordeal, and the ship lands with a pile of bodies pushed into a corner. As in 'Hell Planet', Stone gives her story a high body count and has her survivors learn deadly lessons about the dangers of outer space. Still, 'Human Pets' paints a much rosier and more conservative picture of male space adventures than her work for Lasser.

Despite Stone's nine years of prolific writing, the field was changing in ways that were hostile to her. When she met with new *Astounding* editor John W. Campbell, Jr. about a story in early 1938, he insulted her submission and said, 'I am returning your story, Miss Stone. I do not believe that women are capable of writing science fiction – nor do I approve of it!' (Stone 1997: 101). The rejected story was 'Death Dallies Awhile', which Stone eventually published in the June 1938 issue of *Weird Tales*. Campbell went on to publish dozens of stories by women, including several by C. L. Moore and Leigh Brackett. Clearly, Campbell believed that women could write good SF. What is more likely is that he did not like the critical and 'feminine' type of SF that Stone wrote, a type of SF that could be at home in the pages of *Weird Tales*. The stories Campbell published were decidedly more conservative than those published by Lasser or Tremaine in that they tended to celebrate the colonising possibilities of scientific masculinity to reshape the universe. Stone, whose work always punched holes in the hubris of colonising scientific masculinity, no longer fit within the visions of editors such as Campbell. She published only one more story after 'Death Dallies Awhile' before leaving SF for good.

C. L. Moore's reinvention of frontier adventure

Like Stone, Catherine Lucille Moore grew up with a special love of SF, particularly the work of Edgar Rice Burroughs. In an interview, Moore said of SF, 'I just loved the stories, the fact that they took me out of myself and my narrow little world. I was weaned on the *Mars* books, the *Tarzan* books, and the *Alice in Wonderland* books' (Elliot 1983: 46). Moore attended Indiana University as a literature student, but the economic collapse of the Great Depression forced her to switch to business school so she could get a job at a bank (Yaszek and Sharp 2016: 164). Moore was the 'secretary to the vice-president of the Fletcher Trust Company', a bank in Indianapolis, when she rediscovered SF (Davin 1999: 77). Moore recounts finding an issue of *Amazing Stories* 'at a local newsstand across the street from the bank' during a lunch break (Elliot 1983: 45). She revised a story she began writing as a student and sent it off to Farnsworth Wright at *Weird Tales*, who snapped it up and requested more. Her path to becoming one of the most important SF writers of the mid-twentieth century began with the publication of 'Shambleau' in the November 1933 issue of *Weird Tales*.

Weird Tales was a multi-genre magazine that had a relatively high percentage of women writers, and women such as Moore were able to draw from horror, romance, 'weird' fiction and sword and sorcery when crafting their SF stories (Davin 2006: 65–7). Because of the playful mixing of genres that took place in the magazine, it has routinely been excluded from histories of SF. The most academically powerful argument for such an exclusion came from Darko Suvin, who argued that 'the *fantasy* (ghost, horror, gothic, weird) tale' is 'inimical to the empirical world and its laws', thus making it incompatible with science and SF (2005: 27). Another significant example of this is the important reference work *Science Fiction: The Gernsback Years* by Everett F. Bleiler, which fails to include any stories from *Weird Tales*. Most of C. L. Moore's SF stories from the 1930s were published in *Weird Tales*, so the effect of work such as Bleiler's is to erase many stories by women from the genre's history. 'Tryst in Time', a story Moore published in the December 1936 *Astounding Stories*, had her characteristic complex of genres that included SF, romance, and sword and sorcery. Because of the magazine in which it was published, Bleiler gives a full description of the plot but ends his entry with the note, 'Not really science-fiction' (1998: 300). Such narrow, formalist definitions of SF have tended to exclude or belittle such writing by women in pursuit of a pure, masculinist vision of science and progress for the genre, a vision that extends the sexism of men such as Campbell and celebrates the colonial paradigm most Darwinian feminists and women such as Moore chose to critique. This perspective arrogantly assumes that the weird unknown cannot exist in the same narrative universe with the scientifically known. In excluding the 'weird scientific' SF championed by Farnsworth Wright, too many scholars have diminished our understanding of magazine SF during the Gernsback years and impoverished our understandings of the genre.

'Shambleau' is in many ways the ideal 'weird scientific' story in that it takes a typical space adventure plot and twists it by including elements of horror and fantasy. In particular, 'Shambleau' plays with masculinist and heteronormative expectations about SF and frustrates the standard depiction of the sexy, exotic female alien. It also rejects the essentialist formulations of gender embraced by Darwinian feminists. The story's opening sucks the reader into the tale of morally ambiguous anti-hero Northwest Smith as he ambles through 'the streets of Earth's latest colony on Mars', which is characterised as 'a raw, red little town where anything might happen' (Moore 2016: 166). Smith steps into a doorway and puts 'a wary hand on his heat-gun's grip' when he hears a hysterical mob coming his way shouting 'Shambleau!' (166). The mob is chasing what appears to be 'a berry-brown girl in a single tattered garment' who sees Smith and throws herself at his feet, 'a huddle of burning scarlet and bare, brown limbs' (167). Though Moore's narrator points out that Smith did not have 'the reputation of a chivalrous man', seeing the 'hopeless' girl 'huddle at his feet touched that chord of sympathy for the underdog that stirs in every Earthman' (167). In this way, Moore presents a classic evolutionary 'damsel-in-distress' narrative with the frontier protagonist saving an exotic brown beauty from a mob.

However, from the beginning, Moore troubles this generically safe reading of the story. In an introductory section that precedes the story, Moore draws attention to her promiscuous use of genres by saying that, 'The Myth of Medusa ... can never have had its roots in the soil of Earth. That tale of the snake-haired Gorgon ... never originated about any creature that Earth nourished' (166). Moore's mixture of classical fantasy and space adventure SF leads the reader to doubt the simple appearance of the distressed damsel that the mob calls Shambleau. Once they return to Smith's room, Moore draws upon horror in her revelation to readers how truly alien Shambleau is: while Smith sleeps, he has a 'strange dream' that 'some nameless, unthinkable *thing* ...was coiled about his throat ... something like a soft snake, wet and warm' (174-5; ellipses in original). What is truly unsettling, however, is the mixture of pleasure and disgust that Moore weaves into every sentence. The 'soft snake', which is clearly from the Medusa-like Shambleau, 'was moving gently, very gently, with a soft, caressive pressure that sent little thrills of delight through every nerve and fiber of him, a perilous delight – beyond physical pleasure, deeper than joy of the mind' (175). The 'perilous delight' Smith feels alternates with 'a horror' that turns 'the pleasure into a rapture of revulsion, hateful, horrible – but still most foully sweet' (175). Moore thus queers the standard heterosexual narrative of evolutionary science. Instead of accepting Darwin's formulation of the feminine, Moore creates a distrust for simple gendered appearances through her story. Moore also troubles the usual racial hierarchy of such stories. The 'berry brown' Shambleau is not simply another poor unfortunate from an inferior race who needs to be rescued by a white man. Instead, it is a powerful and complex creature who bends the white man to its will.

The second night, when Shambleau fully reveals itself to Smith while seducing him, Moore destroys the possibility of understanding Shambleau through any stable sense of male/female biological sexual identity. Smith becomes frozen as he sees Shambleau's true form, which is described as 'a nest of blind, restless red worms ... it was – it was like naked entrails endowed with an unnatural aliveness, terrible beyond words' (180; ellipsis in original). In early SF stories, one manifestation of evolutionary essentialism was that women would be overcome by weakness or hysteria when trying to do things outside their supposed domain, like travelling through space or fighting alien men. However, it is the hyper-masculine Smith who falls 'into a blind abyss of submission' (180). Part of the queering of the story comes from this power inversion, with the man becoming the weak one. It is also significant that Smith gets so much pleasure from Shambleau's wet, phallic appendages. Shambleau touches and feeds off of Smith's deepest essence as it 'tickles the very roots of the soul with unnatural delight' (182). Evoking the 'unnatural' biology of the alien calls attention to the standard evolutionary definitions of 'natural' gender that Shambleau defies. Later the soul is referred to as 'the life-forces of men', and Shambleau is described as a creature that has evolved to seduce and feast on men like a 'carnivorous flower' (188). Though Shambleau is referred to as 'she' early in the story, the essence of the creature is androgynous with its true form a pile of phallic wetness that encapsulates Smith while it feeds, and by the end of the story Moore shifts pronouns and refers to Shambleau as an 'it' (186). Though both humans and Venusians – such as Smith's partner Yarol – have males and females in their species, the Shambleau species troubles the common assumption that sexual dimorphism is universal and that white men are superior. Indeed, it is clear that the Shambleau have evolved their pretty feminine guises to trap their prey, thus taking advantage of species that do display sexual dimorphism and racial hierarchies. In this way, Moore uses an evolutionary idea to challenge Darwin's account of sexual selection, race and colonisation that is accepted uncritically in many other SF stories of the period. Shambleau is a creature that is outside the heterosexual economy of sexual selection, and estranges gendered assumptions of masculine and feminine traits that were central to the work of Darwinian feminists.

Moore's gendered performance did not extend to her name. Women writers were a regular feature of *Weird Tales*, and as Moore stated in an interview,

> I wrote 'Shambleau' in the midst of the Depression. The bank was a very paternalistic organization. It was already firing those people whose services weren't really needed. I have the feeling that they might have fired me had they known that I was earning extra income ... Using my initials was simply a means of obscuring my identity. (Elliot 1983: 47)

Wright knew that Moore was a woman, and when fans wrote in letters that mistook her for a man, Wright always took the time to correct them in print. Moore quickly became a fan favourite, publishing a series of Northwest Smith stories. Moore also began a popular series of stories with a groundbreaking Amazon character named Jirel of Joiry. 'Black God's Kiss', the first Jirel story, was published in *Weird Tales* in 1934. The story begins, not surprisingly, with the defeated commander of Joiry brought before the conquering commander Guillaume. 'Joiry's Commander' is described as 'breathing hard' with a 'voice that echoed hollowly under the helmet's confines' that is 'hoarse with fury and despair' (Moore 2007: 21). Guillaume commands his 'men-at-arms' to 'Unshell me this lobster ... We'll see what sort of face the fellow has who gave us such a battle' (21). When they finally succeed in pulling the helmet off, Guillaume utters a 'startled oath' and stares, 'as most men stared when they first set eyes upon Jirel of Joiry. She was tall as most men, and as savage as the wildest of them' (21-2). This first introduction to Jirel invokes both the tradition of the captivity narrative and the trope of the woman in soldier drag. Unlike Darwinian feminists, Moore does not represent her Amazon as a civilised, essentially non-violent heroine. Instead Moore underscores Jirel's 'savage' nature and prowess in battle. Unlike Wilma from the *Buck Rogers* universe, she never faints and does not need to be rescued by men (though men occasionally give her some help). Jirel is the central character in the stories, not a sidekick, and the stories are narrated from her perspective.

Though Moore's stories repeatedly emphasise Jirel's skills as a soldier and leader, most of the narrative time is spent with Jirel going on lonely quests, fighting internal battles or struggling to escape capture. Most of the plots in Moore's Jirel stories follow a revised version of Shakespeare's *Taming of the Shrew*: a strong woman defies men, and then an exceptional man captures her and attempts to tame her. However, Moore's stories have very different endings from Shakespeare's classic play. In 'Black God's Kiss', Guillaume forces a kiss on Jirel, reinforcing the sexual threat he poses to the captured woman soldier. Jirel is outraged, and plans to escape and find a weapon with which to destroy Guillaume. With the help of a priest, she manages to 'escape' into the lowest levels of her dungeons where there is a tunnel that leads to a kind of hell. While this trip to hell uses several elements from fantasy, Moore also provides some scientific details that are consistent with Gernsbackian SF. The tunnel to hell 'leads through poly-dimensional space as well as through the underground – perhaps through time too' (28). When Jirel arrives in this 'hell', she sees 'strange stars' and does not 'recognize a single constellation, and if the brighter sparks were planets they were strange ones, tinged with violet and green and yellow' (32). In effect, this 'hell' is really another planet with less gravity than Earth that is in another dimension occupied by creatures repeatedly described as 'alien' wielding strange technologies. Again, this mixture of fantasy and SF was characteristic of the 'weird scientific' brand of SF Wright favoured for *Weird Tales*.

As Jirel moves through this hell looking for a weapon, she begins to fight the first of many internal battles that she goes through in the stories. Upon seeing a beautiful woman suffering in a mindless frog-like state, Jirel has to overcome her pity and tears to resume the mission. When she returns to this hell in a sequel, she has to fight several battles with an alien who keeps trying to inhabit her body: she uses her violent emotions, bred from her battles and her sordid love life, to fight off the alien's attacks. When Jirel finds the weapon to use against Guillaume in 'Black God's Kiss', it is given to her by the 'black god' of the title: she kisses its statue and becomes possessed by some kind of fatal alien presence. When she returns to the surface, of course she immediately kisses the waiting Guillaume and kills him. Unfortunately, Jirel's violent nature leads her to misread her passionate response to Guillaume's initial kiss. Right after she kills him, she realises 'why such heady violence had flooded her whenever she thought of him ... She knew that there was no light anywhere in the world, now that Guillaume was gone' (50). In other words, Guillaume's violent domination of her released passionate love in her for the first time. This kind of womanly response – where her judgement is overcome by an evolutionary imperative and emotional response – resembles that in Burroughs's Darwinist romances of the early twentieth century. Though a savage warrior woman, Jirel is still a woman who responds to superior might and a forced embrace. She also responds with violence, killing the man who attempts to dominate her. She is a combination of what Darwin asserted were essential masculine and feminine characteristics, a complex character that does not fall back into any standard gender formula.

Jirel is quite different from the swooning damsels of Burroughs or the peaceful Amazons of Darwinian feminism in that her basic emotional response to most things is anger, not fear or compassion. The fact that her anger is stronger than her understanding of love is tragic, but it is also not stereotypical. In the many internal battles she fights – and the many times males try to force a kiss on her – it is always Jirel's anger that allows her to overcome her enemies. Moore's Amazon rejects the feminist invocation of females as essentially cooperative, sympathetic and maternal. She is an independent woman who uses violence to maintain her independence, and provides a strong critique of the evolutionary assumptions of early pulp SF and Darwinian feminism. Finding an anti-essentialist combination of gendered traits was a theme Moore carried into her first publication with Campbell. 'Greater Than Gods' appeared in the July 1939 issue of *Astounding*

Science Fiction, and was the first of several stories Moore published with Campbell over the years. The story dramatises the choices of Dr Bill Cory and how they will affect the future. Bill is a leading scientist in 'Science City' who is working on 'prenatal sex determination' and has mastered the ability to 'separate and identify the genes carrying the factors of sex' and control which sex will be born (Moore 1939: 137). The main choice Bill must make is between which of two women he will marry: one is a gifted chemist Dr Marta Mayhew, and the other is a fun-loving party girl named Sallie Carlisle. As he is meditating on his choice, one of the female descendants created by his choice of Sallie contacts him. This is where Moore provides her clearest rejection of the Darwinian feminist imagination.

Bill's female descendants with Sallie show him what will happen if he chooses her for a wife. In the vision of that possible future, his female descendants eventually establish a matriarchal utopia that leads to happiness and disaster. The utopian pathway begins with women gaining political power:

> Women in public offices were proving very efficient; certainly they governed more peacefully than men. The first woman president won her office on a platform that promised no war so long as a woman dwelt in the White house.
> Of course, some things suffered under the matriarchy. Women as a sex are not scientists, not inventors, not mechanics or engineers or architects. (142)

In this passage, Moore repeats the Darwinian idea that women are not as capable in the sciences as men, an idea that was accepted by some feminists such as Lorraine. However, this future represents a gendered extreme, not a description of what Moore sees as a vital essence of women. This is clear from the fact that Bill's other choice for marriage is a leading chemist, someone whose scientific expertise is a direct contradiction of the future path for humanity associated with the unintelligent Sallie. In other words, this future is a result of Sallie's simplicity, not the simplicity of women as a whole. In this Sallie-oriented future, women gaining control means that science suffers, and the dwindling number of male scientists complain that the world is 'going backward' (142). Their complaints are rejected by the women, who prefer a 'non-mechanized rural civilization' and eventually transform the world into a new Garden of Eden reminiscent of Gilman's *Herland* (142). People began 'slowing mentally and physically', and Bill realises that choosing Sallie would lead to 'a decadent, indolent civilization going down the last incline into oblivion' (143). By portraying peaceful feminine characteristics as leading to decadence, Moore reinforces the Darwinian concept that struggle is necessary for progress. The women of this future are also clearly an expression of Bill's own traits that are brought out by Sallie. As such, this future represents both Sallie's stereotypical characteristics and Bill's 'feminine' side, the part of his character that desires peace and happiness. By making such traits a central aspect of a male character, Moore once again denaturalises the supposed essentials of gender and avoids falling back into formulaic representations of gender.

Moore's critique of Darwinian feminist utopias undoubtedly appealed to Campbell. He also clearly approved of the way Moore's story makes scientific men central to progress. However, Moore also provides a pointed critique of hypermasculine futures that would have been right at home in a Darwinian feminist story. After the first vision of the feminist utopia ceases, a second vision of the future appears to Bill showing him the son he would have with the chemist Marta and the future they would create. This masculine-dominated future is populated by 'men with steel-hard features under steel caps' in a rigidly military society (145). Marta supports his science and brags about his achievements instead of distracting him from his work, pushing him to publish his results quickly instead of taking his time to iron out any problems. As the vision of the masculine future unfolds, Bill learns that his gender-determining procedure causes the children born to have 'a strong strain of obedience, or lack of initiative', and he refuses to use the procedure on Marta (148). The generation of 'followers' that arises quickly support a dictator who plans to take over the world and who uses Bill's procedure to ensure giant families of boys to fight his wars (149). This leads to a 'United World' government, and before long they turn their eyes 'upward toward starry space', conquering Mars and building a violent empire throughout the solar system (150). The 'Leader class' continues to reproduce in the 'old haphazard fashion' and maintains their initiative while requiring all other classes to undergo Bill's procedure, ensuring generations of slavish warriors (151). Moore emphasises the loss of free will and the rise of de facto slavery in this extremely hierarchal society, demonstrating the limitations of colonial scientific masculinity when unfettered from the 'softer' virtues of peace and love.

Moore's representation of these two polarised choices shows the limitation of evolutionary essentialism. Bill has to choose which future he wants to pursue, and neither is particularly appealing until he realises that the descendants he sees in this future are the 'two poles' of his own character drawn out by his two possible choices for a wife. The female future is built on 'his gentler qualities, the tenderness, the deep desire for peace' that Sallie brings out in him, and the male future grows from his 'strong and resolute and proud' characteristics that Marta brings out in him (153). Faced with these two polarised and exaggeratedly gendered choices, Bill chooses a middle path: he asks his assistant Ms Brown to marry him because she won't let him 'neglect the work we're doing' like Sallie, and she won't 'force me to give it to the world unperfected' like Marta (162). By doing so, Moore dramatises the competing impulses within an individual, showing that the feminine and masculine traits of evolutionary essentialism must be present in each individual for them to have balance. In other words, Moore shows the importance of rejecting binary notions of gender and reimagining our common traits as a species. At the same time, Moore rejects the Darwinian feminist imagination by giving a scientific man the power over sexual selection. It is Bill's choice that shapes these futures, and the female characters are merely passive figures who happily mate with him once they are chosen. Moore thus centres the story on a male scientist and his importance for progress, making her story a kind of women's SF that was palatable to Campbell. Appeasing the editorial demands of men such as Campbell became increasingly important to Moore in the 1940s when she decided to attempt SF writing as a full-time occupation. Together with her husband Henry Kuttner, she lived primarily off her SF writing for over a decade, one of the few authors to accomplish such a feat before the 1950s (Yaszek and Sharp 2016: 164).

Leigh Brackett

Moore's contemporary Leigh Brackett also had one of the longest and most distinguished careers among SF writers of the 1940s. She published over sixty-five stories in SF magazines during the course of her forty years in the field, and that is only the beginning of her achievements. Along the way, Brackett published a few novels and became a Hollywood legend for her screenplay writing on such acclaimed films as *The Big Sleep* (1946), *Rio Bravo* (1959) and *Star Wars: The Empire Strikes Back* (1980). Brackett was especially adept at shaping charismatic, tough men who are underdogs in their fights against overwhelming odds. Writing for actors such as Humphrey Bogart, John Wayne and Harrison Ford, Brackett put her stamp on characters such as Philip Marlowe, Sheriff John T. Chance and Han Solo. In essence, Brackett was a superior writer of rough frontier masculinity, whether that frontier was an urban jungle, a besieged town in the American west or the frozen landscape of the ice planet Hoth. She infused her characters with romance and humour, wrapping their core idealism and heroic hearts in thick layers of world-weary swagger and cynicism. In many ways, she was the perfect woman writer for an editor such as John W. Campbell, Jr.

Brackett's first published SF story was 'Martian Quest' in the February 1940 issue of Campbell's *Astounding Science Fiction*. In spite of the antipathy he expressed toward women SF writers, Campbell published a few of Brackett's stories that were space adventures along the lines of what Leslie F. Stone had produced for the past decade. The major difference between Brackett and Stone, however, is that Brackett's stories never questioned the wisdom of colonisation and were always staunchly sympathetic to those working-class men on the ground who did the dirty work of advancing 'civilisation' in the face of extreme 'savagery'. In 'Martian Quest', the heroic man is a young chemist named Martin Drake who arrives in the dry 'Martian Reclaimed Areas' with 'nearly five hundred other passengers' in what is 'their last hope of economic freedom' (Brackett 2002: 1). Martin ends up at the outskirts of a settlement that is about to be shut down because the crops keep getting destroyed by an indigenous lizard species called Khom. He quickly meets and falls in love with a young woman named Terra on a neighbouring farm. Terra is a plucky frontier girl, but her primary role in the story is to spur the hesitant young Martin into action. When Terra learns of Martin's scientific background, she enthuses that when it comes to killing Khom, 'Guns and poison won't do it, but science could find some way, I know it!' (7). Her optimism that science holds the key to successfully settling Mars hearkens back to the most traditional narrative of colonial scientific masculinity. Where most Darwinian feminists would question colonial conquest, Brackett champions the cause of the colonists and portrays the native species of Mars as cruel animals that must give way to the new colonial order.

While Brackett draws on a traditional Darwinist romance plot, she does nothing to reconfigure gender roles. When Martin falters, Terra pushes him to act more like a Darwinian male hero: 'I thought you might still save us, Martin Drake ... Maybe you are a weakling!' (11). Shamed by Terra, Martin finds his courage and uses his scientific skills to ascertain that Khom are really just 'like a termite', and that feeding them acidic melons will destroy the bacteria in their stomachs and eventually starve them (13). Martin triumphantly announces that in 'six months, there won't be a lizard left in the desert!' (15). After a successful last battle against the native lizards, the story ends with Martin having earned the respect of his neighbours and Terra. Armed with his new discovery, it is clear that the frontier town will succeed in killing off the locals and replacing them with new crops, thus transforming the desert. Unlike the space adventure stories of Stone, Brackett celebrates scientific accomplishment and martial bravery as the exclusive domains of men. The role of women is to remind men of their evolutionary duty to save civilisation and subdue nature, thus demonstrating their strength and fitness to survive and reproduce. Campbell clearly felt much more comfortable with this kind of women's writing, which upheld masculinist ideals and did not challenge the wisdom of colonial techno-scientific conquest.

Perhaps the tale that best exemplified how Brackett differed from women such as Lorraine and Stone came with 'The Dragon-Queen of Venus', her eighth published story that appeared in the summer 1941 issue of *Planet Stories* under the editorship of Malcolm Reiss. The story is told from the perspective of Tex, a sergeant in Fort Washington that holds great strategic value in the attempt by Earthmen and Martians to colonise Venus. In a speech to his men, Tex's captain says that,

> Venus is a virgin planet. It's savage, uncivilized, knowing no law but brute force. But it can be built into a great new world. If we do our jobs well, some day these swamps will be drained, the jungles cleared, the natives civilized. The people of Earth and Mars will find new hope and freedom here. (Brackett 2002: 132)

As in 'Martian Quest', Brackett associates the frontier with freedom for the citizens of their home worlds. The desire to colonise Venus is cast in the most noble light, providing a moral justification for the ugly slaughters that Tex and his fellows must commit. Greatly outnumbered, Tex's band must fight off wave after wave of indigenous 'swamp-dwellers' (125). Tex, whose 'ancestor ... died in the Alamo', is associated with the colonial hero who must face off again with an anonymous horde of savages (140). Tex admires his 'savage' adversaries who attack only with 'clubs and crude bows', crediting them because they 'had guts; the kind the Red Indians must have had, back in the old days in America. They had cruelty, too, and a fiendish genius for thinking up tricks' (128). Their tricks include using poisonous snakes and flesh-eating bugs as weapons that they throw into the fort to pick off the soldiers. In this way, Brackett makes explicit the connection between Tex's colonial mission and the United States' wars to take land from Native Americans. No thought is given to whether the natives of Venus have a right to their own land. Like authors such as James Fenimore Cooper, Brackett represents the march of progress as natural and inevitable, with the elimination of the natives a regrettable but necessary part of the march toward civilisation (Sharp 2007: 17-19). To further this narrative, she deploys the stereotype of the depraved savage who is not deserving of civilised treatment.

One twist Brackett gives to the story is the introduction of an Amazon figure, the dragon queen of the title. Brackett's Amazon is identical to the many Amazons produced by her male counterparts: the leader of the Venusian swamp dwellers is just another strong woman who must be tamed or destroyed by heroic males in order for civilisation to be established. Like the Amazons of Greek myth and those of nineteenth-century historian Jakob Bachofen's *Das Mutterrecht* (1861), her power is a clear sign that her people are inherently savage and must be destroyed because of the threat they pose to patriarchal civilisation. In the heat of battle, Tex sees the dragon queen for the first time and is stunned because she 'was beautiful. Pearl-white thighs circling the gray-green barrel of her mount, silver hair streaming from under a snake-skin diadem set with the horns of a swamp-rhino ... Her face was beautiful, too' (129-30). Where Stone's Amazons were alien icons of feminine power who teach lessons to sexist men, Brackett's Amazon is an object of heterosexual male desire who serves to reinforce the need for male power. By standing in the way of progress, the dragon queen actually calls out the required innovation necessary for men to conquer Venus. The fort and its weapons are rusting to nothingness due to the incessant mists and wetness of the environment. The queen infects the fort with mysterious micro-organisms that cause all moisture to evaporate, something that prevents Tex and his Martian sidekick from drinking water. However, Tex figures out how to drink by coating his throat with oil, and learns that the presence of the micro-organisms repels the deadly snakes and bugs that the natives use as weapons. The story concludes with Tex having a eureka moment: 'Scientists could find out how to harness the Dry Spots to keep off the rust, and still let soldiers drink. And some day the swamps would be drained, and men and women would find new wealth, new life, new horizons here on Venus' (144). The dragon queen rides off in fear, seemingly aware that she is doomed to be colonised by Tex and his fellows from Earth and Mars. This happy ending is as far from Darwinian feminism as it is possible to get. Darwinian feminism questioned the evolutionary narrative of masculine control and the supposed superiority of men over women. Brackett imagines worlds where strong women need strong men to subdue them, and that this is the natural order of things.

Women pioneers outside Darwinian feminism

Moore and Brackett were not alone in weaving successful careers in magazine SF without appealing to Darwinian feminism. Amelia Reynolds Long pioneered a style of comic SF that she sold to most of the major editors of the late 1920s and 1930s that anticipated 'the explosion of comic SF' of the 1940s and beyond (Yaszek and Sharp 2016: 213). Her first three stories were for Wright's *Weird Tales* in 1928 and 1930, followed by stories in Gernsback's *Amazing Detective Tales* and *Science Fiction Series*. She also sold stories in the 1930s to magazines such as Sloane's *Amazing Stories*, Tremaine's *Astounding Stories* and William L. Crawford's *Marvel Tales*. Her short story 'Reverse Phylogeny' from the June 1937 issue of *Astounding* shows how women contributed to Tremaine's thought variant style of SF, drawing from folk myths and pseudo-sciences as well as established scientific ideas. Long's playful prose tweaks both scientific masculinity and SF genre conventions through her scientist, an amusing professor figure whose experiments lead to comic chaos instead of suffering, and whose discoveries are only loosely based on the science of the day. 'Reverse Phylogeny' also includes ideas taken from anthropology and archaeology, using the logic of the laboratory experiment to unfold a series of exciting tales from the past in an attempt to establish scientific evidence for the existence of Atlantis. Aside from mocking the image of the male scientist, Long's work bore no resemblance to the Darwinian feminism of Gilman or her contemporaries such as Lorraine and Stone.

Lucile Taylor Hansen was the only woman SF writer to masquerade as a man before the 1940s. Publishing under the name L. Taylor Hansen, she published five stories in *Amazing Stories* and *Wonder Stories* in 1929 and 1930. 'The Man from Space', which she published in the February 1930 issue of *Amazing Stories*, exemplified these changing ideas about SF in the magazine (Hansen 2016). Like Mary Shelley, Charlotte Perkins Gilman and Claire Winger Harris, Hansen used a male narrator to dramatise the possibilities and limitations of patriarchal scientific endeavours. The story focuses on astronomy as witnessed during a fantastic voyage through space. In Sloane's introductory note, he uses his professorial authority to emphasise that the 'author has undoubtedly used good authorities to aid him in his work' (Sloane 1930: 1034). However, Hansen's story begins with a wink at the potentially boring nature of the professorial scientific lectures Sloane was prone to deliver, with the narrator Bob, a college student, nodding off in the middle of a lecture on astronomy. Hansen then draws on the classical story type known as the dream vision, where a guide leads the narrator on a journey of revelation. The guide in Hansen's story is the alien of the title, a silent and ghostly figure who takes Bob, his friend Jim and the bumbling Professor Kepling on a trip to the stars where Bob sees the subject of astronomy come alive. In this sense, Hansen's story exemplified the persistence of Gernsbackian SF's pedagogical mission, using an adventurous journey to teach when scientific lectures cannot. At the same time, Hansen's story exhibits the characteristics Sloane preferred in that she grounded the story in known science and did not present the technical details of space flight as a realistic possibility. Other than poking fun at scientific masculinity, Hansen's fiction had little in common with Darwinian feminism.

Hansen, however, took her tweaking of scientific masculinity to an entirely new level with the aid of *Amazing Stories* editor Ray Palmer in the 1940s. After taking a decade off from SF, during which time she had a daughter and travelled extensively, Hansen returned to write the

'Scientific Mysteries' column for *Amazing*, reporting on recent discoveries for a SF audience (Yaszek and Sharp 2016: 141, 275). Together, Hansen and Palmer portrayed the column as if it was written by a standard, boundaries-pushing example of scientific masculinity. With such column titles as 'The White Race: Does it Exist?' and 'Did the American Indian Once Invade Europe?' Hansen showed a particular penchant for promoting new and controversial theories in anthropology that challenged Euro-American understandings of history and progress. In part this was to encourage readers and SF fans 'to participate in scientific debate on its own terms' (Yaszek and Sharp 2016: 278). It also showed her commitment to understanding and promoting indigenous views of the world, and taking seriously the perspectives of those who had been subjected to the colonial scientific gaze for centuries. Like many of her fellow women of SF, Hansen distrusted the coding of this gaze as masculine and demonstrated in a unique way how women can contribute new insights to both science and SF.

Feminist legacies

Women continued to thrive in SF both during and after the Second World War in spite of the anti-feminist sentiments entrenched in many SF magazines. One place where women played a particularly important role was as editors. Mary Gnaedinger served as the only editor for *Famous Fantastic Mysteries* through its entire run from 1939 to 1953, reprinting the work of many early SF writers such as Minna Irving, Lillian M. Ainsworth and Francis Stevens (Davin 2006: 99–100). In the February 1949 issue, Gnaedinger reprinted Inez Haynes Gillmore's Darwinian feminist novel *Angel Island* in its entirety and introduced it to a new generation of readers. Gnaedinger published new work by men such as Ray Bradbury and women such as Moore, along the way establishing a responsive editorial persona that demonstrated 'sensitivity' to the 'interests and needs' of SF fans (Yaszek and Sharp 2016: 301). At *Weird Tales*, Dorothy McIlwraith took over as editor when Wright's health deteriorated in 1940 (Yaszek and Sharp 2016: 306). McIlwraith continued Wright's welcoming policy toward women writers during her tenure from 1940 to 1954, publishing work by authors such as Mary Elizabeth Counselman, Dorothy Quick and Allison V. Harding (Davin 2006: 132). Both *Famous Fantastic Mysteries* and *Weird Tales* were multi-genre magazines that adopted a broad approach to what counted as SF, so it is no surprise that they were more welcoming to women during this period.

Judith Merril emerged as a major editorial force in the 1950s, compiling 'Year's Best' anthologies beginning in 1956 and becoming the 'main champion' of New Wave SF in the 1960s (Latham 2005: 204). She began as an author of nuclear apocalypse narratives from a uniquely feminine perspective in the late 1940s, drawing on the domestic sphere like Darwinian feminists to provide a strong critique of a scientific masculinity that threatened to destroy the world in the early Cold War. Her first publication, 'That Only a Mother', was precisely this kind of story. Ironically, it appeared in the June 1948 issue of Campbell's *Astounding Science Fiction*. The story centres on a mother who is extremely anxious and in denial about the fact that her child is mutated because of toxic nuclear technologies. Making the central character a hysterical woman instead of a strong Amazon was likely something that made the story palatable to Campbell. However, Merril's story is a domestic update of the battle-of-the-sexes for the nuclear age, where 'the insanity of nuclear war leads to an insane mode of fatherhood based on militaristic thinking' (Yaszek 2008: 115–16). This subtle critique of masculinity became more pronounced in Merril's fiction and the work she promoted as an editor over the course of her career.

The rise of second-wave feminist SF in the late 1960s saw a return of a number of plots popular with Darwinian feminists, but put forward in ways not limited by essentialism, eugenics or heteronormative expectations. Monique Wittig's novel *Les Guérillères* was published in French in 1969 and translated into English in 1971. The novel dramatises a war waged by feminist Amazons against men, and Wittig's Amazons resemble Moore's Jirel of Joiry in that they kill with passion and are not limited by a nurturing essence. As Jane Donawerth notes, 'we can better understand Joanna Russ's female men' when they are put in the context of 'Leslie Stone's cross-dressed hero' Dana Gleason in 'Out of the Void' (1997: 159). Russ wrote her classic lesbian utopia 'When It Changed' in late 1970 and published it in Harlan Ellison's classic New Wave anthology *Again, Dangerous Visions* in 1972 (Russ 1984: 43). The story focuses on a planet of women called Whileaway who have adapted their colony in the aftermath of a plague that killed all the men. They reproduce using new techniques they develop and occupy all roles traditionally occupied by men. The Amazon society of Whileaway does not need men, a fact that is underscored when condescending, sexist men land to restore the colony's contact with Earth and save the women from their supposedly unnatural life. Russ followed this up with the even more unyielding novel *The Female Man* in 1975, which shows similar women in four different worlds. One of the worlds is Whileaway, where it is revealed that the women likely killed the men in a way reminiscent of Gilman's *Herland*. Unlike Gilman, Russ explores the sexual implications of this: in both 'When it Changed' and *The Female Man*, her women have same-sex relationships and build loving families together. The arrival of men is not embraced as a chance to return to a heterosexual economy, but instead is seen as a colonial imposition of an antiquated order that the women strongly resist. Another world is inhabited by Jael, a member of an Amazon race in perpetual war with the men, an aspect of the novel that resonates with stories such as M. F. Rupert's 'Via the Hewitt Ray'.

The legacy of Darwinian feminism is important to reclaim as a formative moment in the long history of women's SF. The myth that women had nothing good to contribute to SF before the 1960s has long been debunked, but still malingers in some academic discussions and in the imaginations of the many fans who have never been exposed to the work of these pioneering women. With the benefit of hindsight, it is easy to see how some Darwinian feminist writing perpetuated regressive ideas about gender, sexuality and race. These failings are hallmarks of the periods in which they were produced, when Euro-American colonial power and scientific masculinity were reshaping the world. Darwinian feminism emerged from within this hegemonic discourse of evolution to address a situation where women had relatively little power over their own bodies, pregnancies were extremely dangerous and education was a privilege few women were allowed. Evolutionary discourse provided a means for women to reinterpret the patriarchal institutions of the past, and gave them rhetorical and narrative tools to imagine how things could evolve in a way that would eliminate the oppressions of the present. Darwinian feminism left its stamp on such popular SF sub-genres as space opera, the techno-utopia and the tale of mad science. It impacted not only women's SF, but also the SF produced by men. The matriarchal Amazons of Darwinian feminism have countless descendants, including 'mother of the future' Sarah Connor from the *Terminator* universe and Lilith from Octavia Butler's *Xenogenesis* series (Sharp 2014). The aviator angels of Darwinian feminism live on in characters such as Kara 'Starbuck' Thrace of the rebooted *Battlestar Galactica* television series and the many women pilots fighting for their countries around the world (Sharp 2010). The idea of scientific femininity informed the 'alternative science' of feminist SF of the 1960s and 1970s (Roberts 1993: 93). It also impacted generations of women scientists and science studies scholars, and has become a staple of ecological discourse. In short, the legacy of Darwinian feminism surrounds us, and it is time for the speculative dimension of Darwinian feminism to be given its rightful place of importance in the history of SF.

Works Cited

Adas, Michael, *Machines as the Measure of Men: Science, Technology, and Ideologies of Western Dominance* (Ithaca: Cornell University Press, 1989).
Aldiss, Brian W., *Trillion Year Spree: The History of Science Fiction* (New York: Avon, 1988).
Anonymous, 'The Female Warrior: An Interesting Narrative of the Sufferings and Singular and Surprising Adventures of Miss Leonora Siddons', in Jesse Alemán and Shelley Streeby (eds), *Empire and the Literature of Sensation: An Anthology of Nineteenth-Century Popular Fiction* (New Brunswick, NJ: Rutgers University Press, 2007), pp. 5–19.
Armitage, David, 'Three Concepts of Atlantic History', in David Armitage and Michael J. Braddick (eds), *The British Atlantic World, 1500–1800* (New York: Palgrave, 2002), pp. 11–27.
Ashley, Mike (ed.), *The Dreaming Sex: Tales of Scientific Wonder and Dread by Victorian Women Writers* (London: Peter Owen, 2010).
_____, 'The Gernsback Days', in Mike Ashley and Robert A. W. Lowndes, *The Gernsback Days: A Study of the Evolution of Modern Science Fiction from 1911 to 1936* (Holicong, PA: Wildside, 2004), pp. 16–254.
_____, *The Time Machines: The Story of the Science-Fiction Pulp Magazines from the Beginning to 1950* (Liverpool: Liverpool University Press, 2000).
Asimov, Isaac (ed.), *Before the Golden Age: A Science Fiction Anthology of the 1930s* (Garden City, NY: Doubleday, 1974).
Attebery, Brian, 'The Conquest of Gernsback: Leslie F. Stone and the Subversion of Science Fiction Tropes', in Justine Larbalestier (ed.), *Daughters of Earth: Feminist Science Fiction in the Twentieth Century* (Middletown, CT: Wesleyan University Press, 2006), pp. 50–66.
_____, *Decoding Gender in Science Fiction* (New York: Routledge, 2002).
Bakhtin, M. M., *Speech Genres & Other Late Essays*, ed. Caryl Emerson and Michael Holquist, trans. Vern W. McGee (Austin: University of Texas Press, 1986).

Bannister, Robert C., *Social Darwinism: Science and Myth in Anglo-American Thought* (Philadelphia: Temple University Press, 1979).

Bazerman, Charles, 'Social Forms as Habitats for Action', *Journal of the Interdisciplinary Crossroads*, 1/2 (August 2004), 123–42.

_____, *Shaping Written Knowledge: The Genre and Activity of the Experimental Article in Science* (Madison: University of Wisconsin Press, 1988).

Bederman, Gail, *Manliness and Civilization: A Cultural History of Gender and Race in the United States, 1880–1917* (Chicago: University of Chicago Press, 1995).

Beebee, Thomas O., *The Ideology of Genre: A Comparative Study of Generic Instability* (University Park, PA: University of Pennsylvania Press, 1994).

Bergman, Jill A., 'The Motherless Child in Pauline Hopkins's *Of One Blood*', *Legacy*, 25/2 (2008), 286–98.

Berkhofer, Robert F. Jr., *The White Man's Indian: Images of the American Indian from Columbus to the Present* (New York: Vintage, 1979).

Blackwell, Antoinette Brown, *The Sexes throughout Nature* (New York: Putnam's, 1875).

Bleiler, Everett F., *Science Fiction: The Gernsback Years* (Kent, OH: Kent State University Press, 1998).

Bourdieu, Pierre, *Language and Symbolic Power*, ed. John B. Thompson, trans. Gino Raymond and Matthew Adamson (Cambridge, MA: Harvard University Press, 1994).

Brackett, Leigh, *Martian Quest: The Early Brackett* (Royal Oak, MI: Haffner Press, 2002).

Brookes, Kristen G., 'A Feminine "Writing that Conquers": Elizabethan Encounters with the New World', *Criticism*, 48/2 (spring 2006), 227–62.

Browne, Janet, *Charles Darwin: The Power of Place* (Princeton, NJ: Princeton University Press, 2002).

_____, *Charles Darwin: Voyaging* (Princeton, NJ: Princeton University Press, 1995).

Bruni, John, *Scientific Americans: The Making of Popular Science and Evolution in Early-Twentieth-Century U.S. Literature and Culture* (Cardiff: University of Wales Press, 2014).

Burroughs, Edgar Rice, *Tarzan of the Apes* (1912; Mineola, NY: Dover, 1997).

Carby, Hazel V., 'Introduction', in Pauline Hopkins, *The Magazine Novels of Pauline Hopkins* (Oxford: Oxford University Press, 1988), pp. xxix–1.

Cavendish, Margaret, *The Blazing World and Other Writings*, ed. Kate Lilley (New York: Penguin, 2004).

Cazden, Elizabeth, *Antoinette Brown Blackwell: A Biography* (Old Westbury, NY: Feminist Press, 1983).

Clark, Ronald W., *The Survival of Charles Darwin: A Biography of a Man and an Idea* (New York: Random House, 1984).

Colley, Linda, 'Perceiving Low Literature: The Captivity Narrative', *Essays in Criticism*, 53/3 (July 2003), 199–218.

Corbett, Elizabeth Burgoyne, *New Amazonia: A Foretaste of the Future*, ed. Alexis Lothian (1889; Seattle: Aquaduct, 2014).

Cottegnies, Line and Nancy Weitz (eds), *Authorial Conquests: Essays on Genre in the Writings of Margaret Cavendish* (Madison: Fairleigh Dickinson University Press, 2003).

Darwin, Charles, *The Voyage of the Beagle: Journal of Researches into the Natural History and Geology of the Countries Visited During the Voyage of H.M.S. Beagle Round the World* (2nd edn, 1845; New York: Modern Library, 2001).

_____, *The Descent of Man* (2nd edn, 1874; Amherst, NY: Prometheus, 1998).

Davin, Eric Leif, *Partners in Wonder: Women and the Birth of Science Fiction, 1926–1965* (Oxford: Lexington, 2006).

_____, *Pioneers of Wonder: Conversations with the Founders of Science Fiction* (Amherst, NY: Prometheus, 1999).

Den Tandt, Christophe, 'Amazons and Androgynes: Overcivilization and the Redefinition of Gender Roles at the Turn of the Century', *American Literary History*, 8/4 (winter 1996), 639–64.

Derounian-Stodola, Kathryn Zabelle (ed.), *Women's Indian Captivity Narratives* (New York: Penguin, 1998).

Desmond, Adrian and James Moore, *Darwin: The Life of a Tormented Evolutionist* (New York: Warner, 1991).

Deutscher, Penelope, 'The Descent of Man and the Evolution of Woman', *Hypatia*, 19/2 (spring 2004), 35–55.

Donawerth, Jane L., 'Illicit Reproduction: Clare Winger Harris's "The Fate of the Poseidonia"', in Justine Larbalestier (ed.), *Daughters of Earth: Feminist Science Fiction in the Twentieth Century* (Middletown, CT: Wesleyan University Press, 2006), pp. 20–35.

_____, *Frankenstein's Daughters: Women Writing Science Fiction* (Syracuse, NY: Syracuse University Press, 1997).

_____, 'Lilith Lorraine: Feminist Socialist Writer in the Pulps', *Science Fiction Studies*, 17 (July 1990), 252–8.

_____ and Carol A. Kolmerten (eds), *Utopian and Science Fiction by Women: Worlds of Difference* (Syracuse, NY: Syracuse University Press, 1994).

Eller, Cynthia, *Gentlemen and Amazons: The Myth of Matriarchal Prehistory, 1861–1900* (Berkeley: University of California Press, 2011).

Elliot, Jeffrey M., 'C. L. Moore: Poet of Far-Distant Futures', *Pulp Voices; or, Science Fiction Voices #6* (San Bernardino, CA: Borgo Press, 1983), pp. 45–51.

Elliott-Baptiste, Marissa, 'Radio-Conscious: The Radio and Early Twentieth-Century Science Fiction' (unpublished conference presentation, Science Fiction Research Association, Riverside, CA, 2013).

Feldman, Mark B., 'Love in the Age of Darwinian Reproduction', in Jeannette Eileen Jones and Patrick B. Sharp (eds), *Darwin in Atlantic Cultures: Evolutionary Visions of Race, Gender, and Sexuality* (New York: Routledge, 2010), pp. 73–89.

Ficke, Sarah H., 'Pauline Hopkins: Rewriting the Imperial Adventure', in Sandra Dixon and Janice Spleth (eds), *Cultural Dynamics of Globalization and African Literature* (Trenton NJ: Red Sea, 2016), pp. 62–77.

Fisher, Lydia, 'American Reform Darwinism Meets Russian Mutual Aid: Utopian Feminism in Mary Bradley Lane's *Mizora*', in Tina Gianquitto and Lydia Fisher (eds), *America's Darwin: Darwinian Theory and U. S. Literary Culture* (Athens, GA: University of Georgia Press, 2014), pp. 181–206.

Freedman, Aviva and Peter Medway (eds), *Genre and the New Rhetoric* (Bristol, PA: Taylor and Francis, 1994).

Gamble, Eliza Burt, *The Evolution of Woman: An Inquiry into the Dogma of her Inferiority to Man* (New York: Knickerbocker, 1894).

Gates, Barbara T. and Ann B. Shteir (eds), *Natural Eloquence: Women Reinscribe Science* (Madison: University of Wisconsin Press, 1997).

Gernsback, Hugo, 'The Wonders of Sleep', *Wonder Stories*, 2/1 (June 1930), 5.

_____, Headnote to 'The Evolutionary Monstrosity', *Amazing Stories Quarterly*, 2/1 (winter 1929), 70.

_____, 'A New Sort of Magazine', *Amazing Stories*, 1/1 (April 1926), 3.
Gillmore, Inez Haynes, *Angel Island* (1914; New York: Plume, 1988).
Gilman, Charlotte Perkins, *The Yellow Wall-Paper, 'Herland', and Selected Writings*, ed. Denise D. Knight (New York: Penguin, 1999).
_____, *Women and Economics: A Study of the Economic Relation Between Men and Women as a Factor in Social Evolution* (1898; Mineola, NY: Dover, 1998).
Godwin, Francis, *The Man in the Moone*, ed. William Poole (1638; Peterborough, ON: Broadview, 2009).
Gould, Stephen Jay, *The Mismeasure of Man* (rev. edn, New York: Norton, 1996).
Gunn, James, 'Introduction', in James Gunn and Matthew Candelaria (eds), *Speculations on Speculation: Theories of Science Fiction* (Lanham, MD: Scarecrow Press, 2005), pp. ix–xi.
_____, 'Toward a Definition of Science Fiction', in James Gunn and Matthew Candelaria (eds), *Speculations on Speculation: Theories of Science Fiction* (Lanham, MD: Scarecrow Press, 2005), pp. 5–12.
_____, *Alternate Worlds: The Illustrated History of Science Fiction* (Englewood Cliffs, NJ: Prentice Hall, 1975).
Hall, Stuart, 'The Whites of Their Eyes: Racist Ideologies and the Media', in Gail Dines and Jean M. Humez (eds), *Gender, Race and Class in Media: A Text-Reader* (Thousand Oaks, CA: Sage, 1995), pp. 18–22.
Hamlin, Kimberly A., *From Eve to Evolution: Darwin, Science, and Women's Rights in Gilded Age America* (Chicago: University of Chicago Press, 2014).
_____, 'The Birds and the Bees: Darwin's Evolutionary Approach to Sexuality', in Jeannette Eileen Jones and Patrick B. Sharp (eds), *Darwin in Atlantic Cultures: Evolutionary Visions of Race, Gender, and Sexuality* (New York: Routledge, 2010), pp. 53–72.
Hansen, L. Taylor, 'The Man from Space', in Lisa Yaszek and Patrick B. Sharp (eds), *Sisters of Tomorrow: The First Women of Science Fiction* (Middletown, CT: Wesleyan University Press, 2016), pp. 144–63.
Haraway, Donna, *Simians, Cyborgs, and Women: The Reinvention of Nature* (New York: Routledge, 1991).
Harding, Sandra, *Is Science Multicultural? Postcolonialisms, Feminisms, and Epistemologies* (Indianapolis: Indiana University Press, 1998).
Harris, Clare Winger, 'The Evolutionary Monstrosity', in Lisa Yaszek and Patrick B. Sharp (eds), *Sisters of Tomorrow: The First Women of*

Science Fiction (Middletown, CT: Wesleyan University Press, 2016), pp. 9–25.

  ———, *Away from the Here and Now: Stories in Pseudo-Science* (1947; San Bernardino, CA: Surinam Turtle Press, 2011).

  ———, 'The Fate of the Poseidonia', in Justine Larbalestier (ed.), *Daughters of Earth: Feminist Science Fiction in the Twentieth Century* (Middletown, CT: Wesleyan University Press, 2006), pp. 1–18.

Hausman, Bernice L., 'Sex before Gender: Charlotte Perkins Gilman and the Evolutionary Paradigm of Utopia', *Feminist Studies*, 24/3 (fall 1998), 489–510.

Herodotus, 'The Amazons', in Sam Moskowitz (ed.), *When Women Rule* (New York: Walker, 1972), pp. 28–30.

Hopkins, Pauline, *The Magazine Novels of Pauline Hopkins* (Oxford: Oxford University Press, 1988).

Hubbard, Ruth, 'Have Only Men Evolved?', in Janet A. Kourany (ed.), *The Gender of Science* (Upper Saddle River, NJ: Prentice, 2002), pp. 153–70.

Hudson, Nicholas, 'From "Nation" to "Race": The Origin of Racial Classification in the Eighteenth-Century Thought', *Eighteenth-Century Studies*, 29/3 (spring 1996), 247–64.

Hutton, Sarah, 'Science and Satire: The Lucianic Voice of Margaret Cavendish's *Description of a New World Called the Blazing World*', in Line Cottegnies and Nancy Weitz (eds), *Authorial Conquests: Essays on Genre in the Writings of Margaret Cavendish* (Madison: Fairleigh Dickinson University Press, 2003), pp. 161–78.

Irving, Minna, 'The Moon Woman', *Amazing Stories*, 4/8 (November 1929), 746–54.

James, Edward, *Science Fiction in the Twentieth Century* (New York: Oxford University Press, 1994).

Jann, Rosemary, 'Revising the Descent of Woman: Eliza Burt Gamble', in Barbara T. Gates and Ann B. Shteir (eds), *Natural Eloquence: Women Reinscribe Science* (Madison: University of Wisconsin Press, 1997), pp. 147–63.

Jones, Margaret C., *Heretics and Hellraisers: Women Contributors to 'The Masses', 1911–1917* (Austin: University of Texas Press, 1993).

Kaplan, Amy, 'Manifest Domesticity', *American Literature*, 70/3 (September 1998), 581–606.

Kaplan, Gisela and Lesley J. Rogers, *Gene Worship: Moving Beyond the Nature/Nurture Debate over Genes, Brain, and Gender* (New York: Other Press, 2003).

Keller, Evelyn Fox, *Reflections on Gender and Science* (New Haven: Yale University Press, 1985).

Ketterer, David, *New Worlds for Old: The Apocalyptic Imagination, Science Fiction, and American Literature* (Bloomington: Indiana University Press, 1974).

Khanna, Lee Cullen, 'The Subject of Utopia: Margaret Cavendish and Her *Blazing-World*', in Jane L. Donawerth and Carol A. Kolmerten (eds), *Utopian and Science Fiction by Women: Worlds of Difference* (Syracuse, NY: Syracuse University Press, 1994), pp. 15–34.

Kincaid, Paul, 'On the Origins of Genre', *Extrapolation*, 44/4 (winter, 2003), 409–19.

Kolmerten, Carol, 'Texts and Contexts: American Women Envision Utopia, 1890–1920', in Jane L. Donawerth and Carol A. Kolmerten (eds), *Utopian and Science Fiction by Women: Worlds of Difference* (Syracuse, NY: Syracuse University Press, 1994), pp. 107–25.

Kuhn, Thomas, *The Structure of Scientific Revolutions* (2nd edn, Chicago: University of Chicago Press, 1970).

Lane, Mary E. Bradley, *Mizora: A World of Women* (1890; Lincoln: University of Nebraska Press, 1999).

Larbalestier, Justine (ed.), *Daughters of Earth: Feminist Science Fiction in the Twentieth Century* (Middletown, CT: Wesleyan University Press, 2006).

_____, *The Battle of the Sexes in Science Fiction* (Middletown, CT: Wesleyan University Press, 2002).

Latham, Rob, 'The New Wave', in David Seed (ed.), *A Companion to Science Fiction* (Malden, MA: Blackwell, 2005), pp. 201–16.

Long, Amelia Reynolds, 'Reverse Phylogeny', in Lisa Yaszek and Patrick B. Sharp (eds), *Sisters of Tomorrow: The First Women of Science Fiction* (Middletown, CT: Wesleyan University Press, 2016), 213–22.

Lorraine, Lilith, 'Letter to *Acolyte*, Spring 1943', in Lisa Yaszek and Patrick B. Sharp (eds), *Sisters of Tomorrow: The First Women of Science Fiction* (Middletown, CT: Wesleyan University Press, 2016), p. 315.

_____, 'Into the 28th Century', in Lisa Yaszek and Patrick B. Sharp (eds), *Sisters of Tomorrow: The First Women of Science Fiction* (Middletown, CT: Wesleyan University Press, 2016), pp. 108–41.

———, 'The Celestial Visitor', *Wonder Stories*, 6/10 (March 1935), 1190-207.
———, 'The Isle of Madness', *Wonder Stories*, 7/6 (November-December 1935), 653-67.
———, 'The Jovian Jest', *Astounding Stories*, 2/2 (May 1930), 228-33.
———, 'The Brain of the Planet', *Science Fiction Series*, 5 (1929), 2-23.
Loewen, James W., *Lies My Teacher Told Me: Everything Your American History Textbook Got Wrong* (rev edn, New York: Touchstone, 2007).
Luckhurst, Roger, 'Bruno Latour's Scientifiction: Networks, Assemblages, and Tangled Objects', *Science Fiction Studies*, 33/1 (March 2006), 4-17.
———, 'Introduction', *Science Fiction Studies*, 33/1 (March 2006), 1-3.
———, *Science Fiction* (Malden, MA: Polity, 2005).
Luckmann, Thomas, 'On the Communicative Adjustment of Perspectives, Dialogue and Communicative Genres', in Astri Heen Wold (ed.), *The Dialogue Alternative* (Oslo: Scandinavian University Press, 1992), pp. 219-34.
MacLeod, Roy, 'Introduction', *Osiris*, 15 (2000), 1-13.
Mather, Cotton, 'A Notable Exploit; wherein, *Dux Faemina Facti* from *Magnalia Christi Americana*', in Kathryn Zabelle Derounian-Stodola (ed.), *Women's Indian Captivity Narratives* (New York: Penguin, 1998), pp. 58-60.
Merchant, Carolyn, '"The Violence of Impediments": Francis Bacon and the Origins of Experimentation', *Isis*, 99/4 (December 2008), 731-60.
———, 'The Scientific Revolution and the Death of Nature', *Isis*, 97/3 (September 2006), 513-33.
———, *The Death of Nature: Women, Ecology, and the Scientific Revolution* (1980; New York: HarperCollins, 1990).
Milam, Erika Lorraine and Robert A. Nye, 'An Introduction to *Scientific Masculinities*', *Osiris*, 30 (2015), 1-14.
Milburn, Colin, *Mondo Nano: Fun and Games in the World of Digital Matter* (Durham, NC: Duke University Press, 2015).
Miller, Carolyn R., 'Genre as Social Action', in Aviva Freedman and Peter Medway (eds), *Genre and the New Rhetoric* (Bristol, PA: Taylor and Francis, 1994), pp. 23-42.
Montrose, Louis, 'The Work of Gender in the Discourse of Discovery', *Representations*, 33 (1991), 1-41.
Moore, C. L., 'Shambleau', in Lisa Yaszek and Patrick B. Sharp (eds), *Sisters of Tomorrow: The First Women of Science Fiction* (Middletown, CT: Wesleyan University Press, 2016), pp. 166-90.

———, *Black God's Kiss* (Bellevue, WA: Planet Stories, 2007).
———, 'Greater Than Gods', *Astounding Science Fiction*, 23/5 (July 1939), 135–62.
Moskowitz, Sam (ed.), *When Women Rule* (New York: Walker, 1972).
Murphy, Patricia, *In Science's Shadow: Literary Constructions of Late Victorian Women* (Columbia, MO: University of Missouri Press, 2006).
Namias, June, *White Captives: Gender and Ethnicity on the American Frontier* (Chapel Hill: University of North Carolina Press, 1993).
Noble, David F., *A World Without Women: The Christian Clerical Culture of Western Science* (New York: Knopf, 1992).
Nott, Josiah C., 'Geographical Distribution of Animals, and the Races of Men', in Josiah C. Nott and George R. Gliddon, *Types of Mankind* (Philadelphia: Lippincott Grambo, 1854), pp. 62–80.
Nowlan, Philip Francis, 'The Airlords of Han', *Amazing Stories*, 3/12 (March 1929), 1108–36.
———, 'Armageddon 2419 A. D.', *Amazing Stories*, 3/5 (August 1928), 422–49.
Park, Katherine, 'Women, Gender, and Utopia: *The Death of Nature* and the Historiography of Early Modern Science', *Isis*, 97.3 (September 2006), 487–95.
Paxton, Nancy L., *George Eliot and Herbert Spencer: Feminism, Evolutionism, and the Reconstruction of Gender* (Princeton, NJ: Princeton University Press, 1991).
Penley, Constance, *NASA/TREK: Popular Science and Sex in America* (New York: Verso, 1997).
Poole, William, 'Preface', in Francis Godwin, *The Man in the Moone*, ed. William Poole (1638; Peterborough, ON: Broadview, 2009), pp. 7–8.
Rich, Charlotte, 'From Near-Dystopia to Utopia: A Source for *Herland* in Inez Haynes Gillmore's *Angel Island*', in Cynthia J. Davis and Denise D. Knight (eds), *Charlotte Perkins Gilman and Her Contemporaries: Literary and Intellectual Contexts* (Tuskaloosa, AL: University of Alabama Press, 2004), pp. 171–93.
Rieder, John, *Science Fiction and the Mass Cultural Genre System* (Middletown, CN: Wesleyan University Press, 2017).
———, *Colonialism and the Emergence of Science Fiction* (Middletown, CN: Wesleyan University Press, 2008).
Roberts, Robin, *A New Species: Gender and Science in Science Fiction* (Urbana: University of Illinois Press, 1993).

Robinson, Michael, 'Manliness and Exploration: The Discovery of the North Pole', *Osiris*, 30 (2015), 89–109.

Rosen, Christine, *Preaching Eugenics: Religious Leaders and the American Eugenics Movement* (New York: Oxford University Press, 2004).

Rossiter, Margaret W., *Women Scientists in America: Struggles and Strategies to 1940* (Baltimore: Johns Hopkins University Press, 1982).

Roughgarden, Joan, *Evolution's Rainbow: Diversity, Gender, and Sexuality in Nature and People* (Berkeley: University of California Press, 2004).

Rowlandson, Mary, 'A True History of the Captivity and Restoration of Mrs. Mary Rowlandson', in Kathryn Zabelle Derounian-Stodola (ed.), *Women's Indian Captivity Narratives* (New York: Penguin, 1998), pp. 7–51.

Rupert, M. F., 'Via the Hewitt Ray', *Science Wonder Quarterly*, 1/3 (spring 1930), 370–83, 420.

Russ, Joanna, 'When It Changed', in Jeffrey M. Elliot (ed.), *Kindred Spirits: An Anthology of Gay and Lesbian Science Fiction Stories* (Boston: Alyson, 1984), pp. 43–53.

Sarasohn, Lisa T., *The Natural Philosophy of Margaret Cavendish: Reason and Fancy during the Scientific Revolution* (Baltimore: Johns Hopkins University Press, 2010).

Sargent, Pamela, *Women of Wonder: Science Fiction Stories by Women about Women* (New York: Vintage, 1975).

Satter, Beryl, *Each Mind a Kingdom: American Women, Sexual Purity, and the New Thought Movement, 1875–1920* (Berkeley: University of California Press, 1999).

Schiebinger, Londa, 'Feminist History of Colonial Science', *Hypatia*, 19/1 (February 2004a), 233–54.

———, *Nature's Body: Gender in the Making of Modern Science* (1993; New Brunswick, NJ: Rutgers University Press, 2004b).

———, *The Mind Has No Sex? Women in the Origins of Modern Science* (Cambridge, MA: Harvard University Press, 1989).

Shapin, Steven, *A Social History of Truth: Civility and Science in Seventeenth-Century England* (Chicago: University of Chicago Press, 1994).

Sharp, Patrick B., 'Darwinism', in Rob Latham (ed.), *The Oxford Handbook of Science Fiction* (New York: Oxford University Press, 2014), pp. 475–85.

_____, 'The Evolution of the West: Darwinist Visions of Race and Progress in Roosevelt and Turner', in Jeannette Eileen Jones and Patrick B. Sharp (eds), *Darwin in Atlantic Cultures: Evolutionary Visions of Race, Gender, and Sexuality* (New York: Routledge, 2010), pp. 225-36.

_____, 'Starbuck as "American Amazon": Captivity Narrative and the Colonial Imagination in *Battlestar Galactica*', *Science Fiction Film and Television* 3.1 (2010), 57-78.

_____, *Savage Perils: Racial Frontiers and Nuclear Apocalypse in American Culture* (Norman: University of Oklahoma Press, 2007).

Shelley, Mary, *Frankenstein* (rev. edn, 1831; New York: Barnes & Noble, 2003).

Sloane, T. O'Conor, Headnote to 'The Man from Space', *Amazing Stories*, 4/11 (February 1930), 1034.

_____, 'Acceleration in Interplanetary Travel', *Amazing Stories*, 4/8 (November 1929), 677.

Slocum, Sally, 'Woman the Gatherer: Male Bias in Anthropology', in Rayna Rapp Reiter (ed.), *Toward an Anthropology of Women* (New York: Monthly Review Press, 1975), pp. 36-50.

Slotkin, Richard, *Regeneration Through Violence: The Mythology of the American Frontier, 1600-1860* (1973; New York: HarperPerennial, 1996).

Spencer, Herbert, 'Psychology of the Sexes', *Popular Science Monthly*, 4 (1873; New York: Appleton, 1874), pp. 30-8.

Stone, Leslie F., 'Out of the Void', in Lisa Yaszek and Patrick B. Sharp (eds), *Sisters of Tomorrow: The First Women of Science Fiction* (Middletown, CT: Wesleyan University Press, 2016), pp. 27-105.

_____, 'The Conquest of Gola', in Justine Larbalestier (ed.), *Daughters of Earth: Feminist Science Fiction in the Twentieth Century* (Middletown, CT: Wesleyan University Press, 2006), pp. 36-49.

_____, 'Day of the Pulps', *Fantasy Commentator* (fall 1997), 100-3, 152.

_____, 'The Human Pets of Mars', in Isaac Asimov (ed.), *Before the Golden Age: A Science Fiction Anthology of the 1930s* (Garden City, NY: Doubleday, 1974), pp. 729-75.

_____, 'The Fall of Mercury', *Amazing Stories*, 10/7 (December 1935), 27-73.

_____, 'The Hell Planet', *Wonder Stories*, 4/1 (June 1932), 14-27.

_____, 'Through the Veil', *Amazing Stories*, 5/2 (May 1930), 174-82.

———, 'Women with Wings', *Air Wonder Stories*, 1/11 (May 1930), 984–1003.

———, 'A Letter of the Twenty-Fourth Century', *Amazing Stories*, 4/9 (December 1929), 860–1.

———, 'Men with Wings', *Air Wonder Stories*, 1/1 (July 1929), 58–87.

Suvin, Darko, 'Estrangement and Cognition', in James Gunn and Matthew Candelaria (eds), *Speculations on Speculation: Theories of Science Fiction* (Lanham, MD: Scarecrow Press, 2005), pp. 23–35.

———, *Metamorphoses of Science Fiction: On the Poetics and History of a Literary Genre* (New Haven, CT: Yale University Press, 1979).

Terrall, Mary, 'Heroic Narratives of Quest and Discovery', *Configurations*, 6/2 (spring 1998), 223–42.

Thomas, David Hurst, *Skull Wars: Kennewick Man, Archaeology, and the Battle for Native American Identity* (New York: Basic, 2000).

Tinnemeyer, Andrea, *Identity Politics of the Captivity Narrative after 1848* (Lincoln: University of Nebraska Press, 2006).

Todorov, Tzvetan, *Genres in Discourse*, trans. Catherine Porter (New York: Cambridge University Press, 1990).

Venet, Gisele, 'Margaret Cavendish's Drama: An Aesthetic of Fragmentation', in Line Cottegnies and Nancy Weitz (eds), *Authorial Conquests: Essays on Genre in the Writings of Margaret Cavendish* (Madison: Fairleigh Dickinson University Press, 2003), pp. 213–28.

Vint, Sherryl, *Bodies of Tomorrow: Technology, Subjectivity, Science Fiction* (Toronto: University of Toronto Press, 2007).

'Wallin's Theory of Evolution', *Amazing Stories Quarterly*, 2/1 (winter 1929), 73.

Walters, Lisa, *Margaret Cavendish: Gender, Science and Politics* (Cambridge: Cambridge University Press, 2014).

Ware, Susan, *Still Missing: Amelia Earhart and the Search for Modern Feminism* (New York: W. W. Norton, 1993).

Weinbaum, Batya, *Islands of Women and Amazon: Representations and Realities* (Austin: University of Texas Press, 1999).

Weinberg, Robert, *The Weird Tales Story* (1977; Berkeley Heights, NJ: Wildside, 1999).

Westfahl, Gary, *Hugo Gernsback and the Century of Science Fiction* (Jefferson, NC: McFarland, 2007).

———, *The Mechanics of Wonder: The Creation of the Idea of Science Fiction* (Liverpool: Liverpool University Press, 1998).

Williams, Nathaniel, 'Reconstructing Biblical History: Garrett Serviss, Pauline Hopkins, and Technocratic Exploration Novels', *Nineteenth-Century Contexts*, 34/4 (September 2012), 323–40.

Wosk, Julie, *Women and the Machine: Representations from the Spinning Wheel to the Electronic Age* (Baltimore: Johns Hopkins University Press, 2001).

Yaszek, Lisa, *Galactic Suburbia: Recovering Women's Science Fiction* (Columbus, OH: Ohio State University Press, 2008).

―――― and Patrick B. Sharp (eds), *Sisters of Tomorrow: The First Women of Science Fiction* (Middletown, CT: Wesleyan University Press, 2016).

Zinn, Howard, *A People's History of the United States, 1492–Present* (New York: HarperPerennial, 2005).

Index

A

Air Wonder Stories, 117, 121, 130, 131
Amazing Stories, 19, 99, 101, 103, 105, 108, 118, 121, 126, 144–7, 166, 169–70
Amazonian flower, 76–7, 103–4
Amazons, 11, 24–5, 69–79, 83–5, 89–90, 92, 94–5, 103–5, 108, 116–17, 119, 122–4, 126, 132–3, 140, 148, 160–2, 168, 172–3
American Amazon, 72–7, 79, 103, 108, 116–17
angels, 11, 36, 52, 98–104, 117–20, 133–4, 171–4
Asimov, Isaac, 155–6
Astounding Stories / Science Fiction, 147–8, 153, 156, 157, 162, 166, 169, 172

B

Bachofen, Johann Jakob, 61, 69, 168
Bacon, Francis, 12, 15, 17, 19–20, 22, 24, 26, 27, 30, 33, 37, 38, 39
  The New Atlantis, 15, 19–20
Bates, Harry, 148
birth control, 9–10, 64, 70, 95, 98, 151–2
Blackwell, Antoinette Brown, 6, 35, 55–61, 66, 67, 77, 79–83, 90, 93, 97, 101, 125
  The Sexes Throughout Nature, 56–61
Boyle, Robert, 19, 21–2, 33
Brackett, Leigh, 148, 156, 165–9
  'Martian Quest', 166–7
  'The Dragon-Queen of Venus', 167–9
Burroughs, Edgar Rice, 7, 16, 86, 88, 93, 102, 114, 119, 126–7, 134–5, 137, 157, 162
  A Princess of Mars, 86, 126
  Tarzan of the Apes, 86, 114, 157

C

Campbell, John W. Jr., 156, 158, 162–7, 172
captivity narrative, 71–7, 89, 93, 103, 108, 161
Cavendish, Margaret, 1, 2, 8, 20–6, 31, 34, 120, 149
  The Blazing World, 1, 21–6, 149
childbirth, 9–10, 33, 64, 72, 92, 95, 98, 101, 130–2, 134, 141–2, 151–2
civilisation, 7, 10, 15–16, 30–1, 35–6, 40–2, 46, 50, 52, 55, 58, 61–2, 69, 71, 74–5, 79, 81, 84–8, 93, 96, 98, 102, 109, 119–21, 124–5, 129, 139, 141, 145, 150–3, 161, 166–8
clerical asceticism, 19
Columbus, Christopher, 14–5, 22, 27, 42, 154

colonial gaze, 10, 18, 41–3, 102, 134–5, 150, 171
colonisation, 1, 10, 13, 15–18, 22, 28–30, 33, 35, 42–3, 46, 70–2, 78, 83–5, 94, 98, 102, 120, 129–30, 134–6, 154–6, 160–1, 167–8
Corbett, Elizabeth Burgoyne, 83–5
New Amazonia, 83–5

D
Darwin, Charles, 1–2, 6–8, 10, 12, 34–55, 57–60, 62–4, 66–7, 93–6, 105, 109, 117, 125, 133, 140, 142, 159–60
The Voyage of the Beagle, 36–43, 52
The Descent of Man, 6, 10, 34, 43–52, 54–5, 62, 66, 93, 109
discovery narrative, 14–17, 23, 26–9, 32, 38–9, 42–5, 51, 71, 78–9, 81, 85–6, 88, 92, 94, 118–19, 153–4, 171
domestic sphere, 9–11, 29, 40, 51–2, 55, 60, 67, 70–2, 75, 80–5, 89, 93, 96, 101–2, 106, 110, 112, 114, 117, 119–20, 136–7, 142
Donawerth, Jane, L., 1, 9, 101, 107, 172
Dunstan, Hannah, 72–8

E
Earhart, Amelia, 11, 117–18, 122–3, 126, 146–7, 157–8, 173
education, 5, 9, 54–5, 61, 65–7, 77, 80, 96, 101, 126, 149, 173
Ellis, Sophie Wenzel, 148

eugenics, 11, 17, 64–5, 82–4, 92, 96–7, 124–5, 131–2, 141, 143, 152, 172
evolutionary essentialism, 34, 55, 58, 65, 89, 103, 124, 143, 159, 165

F
Famous Fantastic Mysteries, 87, 118, 171
The Female Warrior (anon.), 75–7

G
Gamble, Eliza Burt, 6, 56, 61–7, 69–70, 79, 83, 88, 90, 93, 95, 97, 101, 121, 122, 141, 142, 151
The Evolution of Woman, 61–7, 69–70, 141
genre, 1–15, 21, 26–8, 31, 33, 51, 71–5, 101–2
Gernsback, Hugo, 1, 99, 101–5, 108, 112, 117–18, 121–2, 125, 130, 134, 137, 144–9, 151, 153, 157–8, 161, 169–70
Gillmore, Inez Haynes, 6, 7, 86–93, 97, 119–20, 131, 134, 141, 142, 171
Angels and Amazons, 90
Angel Island, 86–92, 119–20, 134, 141, 171
Gilman, Charlotte Perkins, 2, 6, 7, 10, 67, 87–99, 101–2, 113, 122, 124, 130–1, 133, 138, 140–3, 148, 151, 164, 170, 173
Herland, 10, 67, 87, 93–9, 102, 113, 122, 124, 130, 133, 140–1, 164, 173

Women and Economics, 67,
    86–7, 93
'The Yellow Wall Paper', 89
Gnaedinger, Mary, 87, 171
Godwin, Francis, 15–17, 23, 120
    The Man in the Moone, 15–7, 23

H
Hamlin, Kimberly A., 6, 43, 60, 101
Hansen, L(ucile) Taylor, 34, 61,
    146, 147, 170–1
    'The Man from Space', 170
Harris, Clare Winger, 2, 7–8, 34,
    61, 102, 105–17, 120, 138–9,
    143, 146, 147, 150, 170
    'The Ape Cycle', 112–17
    'The Evolutionary Monstrosity',
        108–12, 139
    'The Fate of the Poseidonia',
        107–8
    'A Runaway World', 105–7
Herodotus, 70–1
heterosexual economy of desire,
    79, 95, 98, 120, 122, 125,
    133–4, 148, 159–60, 168, 173
Hopkins, Pauline, 85–6
    Of One Blood, 85–6
Hornig, Charles D., 149, 151, 153

I
Irving, Minna, 117–21, 134, 142,
    146, 171
    'The Moon Woman', 117–21,
        134, 142

L
Lane, Mary E. Bradley, 2, 77–85,
    87, 94, 104, 121, 138
    Mizora, 77–83, 85, 87, 94

Lasser, David, 121–2, 125, 130,
    134, 137, 140, 144, 148,
    153–6
Long, Amelia Reynolds, 34, 121,
    137, 146, 148–9, 169–70
    'Reverse Phylogeny', 169–70
Lorraine, Lilith, 10, 34, 61, 82, 102,
    121, 131, 137–44, 148–53,
    163, 167, 170
    'The Brain of the Planet',
        137–40
    'The Celestial Visitor', 149–51
    'Into the 28th Century', 140–4
    'The Isle of Madness', 151–3
    'The Jovian Jest', 148–9
    'Letter to *Acolyte*', 153
Luckhurst, Roger, 5, 13

M
mad science, 11, 31–4, 102,
    108–10, 112–15, 139, 173
man the toolmaker, 7, 49–51, 58,
    94, 105, 140
marriage, 6, 36, 61, 64, 77, 88–9,
    91–2, 94, 97, 111–12, 114,
    116, 120, 129, 143, 150–1,
    163
Mars, 86, 106–8, 126–7, 134–6,
    155, 157–8, 164, 166–7,
    169
Mather, Cotton, 72–5
McIlwraith, Dorothy, 171
Mechanism, 18, 21, 31, 33
Merchant, Carolyn, 18, 20
Merril, Judith, 106, 172
Moore, C(atherine) L(ucille), 2,
    7–8, 120, 148, 156–65, 169,
    171–2
    'Black God's Kiss', 160–2
    'Greater Than Gods', 163–5
    'Shambleau', 120, 157–60

motherhood, 33, 36, 40, 51–4, 58–62, 64–5, 67, 72, 74–5, 79–84, 86, 95–6, 103, 115, 120–1, 125, 134, 141, 172–3

N

Newton, Isaac, 26–7, 39
Nott, Josiah C., 115
Nowlan, Philip Francis, 103–5
 'The Airlords of Han', 103–5
 'Armageddon 2419 A. D.', 103–5

O

overcivilisation, 115–16

P

Palmer, Ray, 170–1
Planet Stories, 167

R

race, 6–7, 14, 29–31, 34–5, 39, 43–4, 46, 49–51, 54, 62, 67, 73, 77, 82, 85–6, 88, 91–2, 98, 102–3, 107–8, 113–15, 120, 128–32, 139, 141, 143, 152, 159–60, 171, 173
rape, 10, 70, 76, 88, 94–5, 114, 116
Reiss, Malcolm, 167
reproduction, 70, 79, 82–5, 92, 95–8, 101, 111–12, 117, 124, 133–4, 141–2, 151–2, 164, 167, 172
rhetoric of certainty, 44–5, 50, 54
Rieder, John, 5–6, 10, 33
Rowlandson, Mary, 72–3
Royal Society of London, 19–22, 24, 26, 38

Rupert, M. F., 121–6, 131, 133, 173
 'Via the Hewitt Ray', 122–6, 173
Russ, Joanna, 74, 172–3
 'When it Changed', 172–3

S

Sanger, Margaret, 98–9, 151
savagery, 7, 10–11, 16, 23, 30–1, 36, 40–2, 46, 48, 51–2, 54, 61, 71–4, 77–8, 94, 96, 102, 107–9, 120–1, 129, 132, 136, 139, 145, 150–4, 161–2, 166–8
Science Fiction Series, 121, 137, 169
Science Wonder Quarterly, 112, 121–2, 140
Science Wonder Stories, 121
scientific femininity, 6, 11, 34–5, 56–61, 65–6, 77, 80–3, 85, 95–6, 99, 112, 117, 119, 123–5, 128, 136–7, 174
scientific masculinity, 6, 9, 11, 13–22, 26–35, 38–41, 46, 50, 57–8, 93–4, 98, 109–10, 112–14, 118, 123, 126, 131, 135, 139, 149, 153, 156, 164–6, 169–73
selfishness, 51–2, 55, 59, 61–3, 112, 114
sexual desire, 58, 63–4, 79, 83, 90–2, 95, 97, 125, 141–2, 160, 168
sexual selection, 2, 6, 9–10, 43–52, 55, 58, 63–5, 67, 77, 79–80, 83, 88, 92–3, 95, 97–8, 102, 105, 112, 116, 122, 124–5, 127, 131, 133–5, 140–2, 150–1, 160, 165

Shelley, Mary, 2, 3, 8, 11, 26, 31–4, 92–3, 102, 109–10, 123, 148, 170
   Frankenstein, 31–4, 109–10, 124, 133
Sloane, T. O'Conor, 118, 121, 130, 146–7, 154–5, 169–70
Spencer, Herbert, 36, 45, 52–5, 57–60, 66, 110
   'Psychology of the Sexes', 53–4
Stone, Leslie F., 2, 7–8, 34, 102, 121, 125–37, 141–4, 146, 148, 153–6, 166–8, 170, 172
   'The Conquest of Gola', 134–7, 153–4
   'The Hell Planet', 153–5
   'The Human Pets of Mars', 155–6
   'A Letter of the Twenty-Fourth Century', 129–30
   'Men with Wings', 130–1
   'Out of the Void', 126–9, 154, 172
   'Through the Veil', 129–30
   'Women with Wings', 131–4, 136, 141–2
sympathy, 10, 52, 62, 78, 91

T
travelogue, 6, 15–17, 23, 28–31, 38–9, 51–2, 77–8

U
utopia, 3, 8, 10, 12, 15–16, 20, 22–4, 26, 67, 77–9, 82–8, 91–6, 112–13, 120, 122–4, 129–33, 138–43, 148–52, 163–4, 172–3

V
Venus, 106, 126, 132–7, 140, 150, 160, 167–9
Verne, Jules, 16, 28, 88
violence, 2, 6–7, 15, 25, 48, 51–2, 55, 61, 69–70, 73–5, 78–9, 83, 85–6, 89, 93–6, 98, 103, 112, 114, 116, 119–20, 122–4, 135, 138–40, 154, 162, 164

W
Weird Tales, 101, 105, 118, 146–8, 156–7, 160–1, 169, 171
Wells, H. G., 1, 34, 66, 88, 102, 109–10, 112, 123, 138, 155
Wittig, Monique, 172
women's suffrage, 2, 6, 34, 56, 62, 66, 83–4, 87, 92, 98
Wonder Stories, 7, 99, 101, 121, 134, 144, 149, 151, 153, 170
Wright, Farnsworth, 146–7, 157–8, 160–1, 169, 171